ROGUE PRESS

PRESENTS

THE "GOD" PART
OF THE BRAIN

A SCIENTIFIC
INTERPRETATION OF
HUMAN SPIRITUALITY
AND GOD

MATTHEW ALPER

Published by Rogue Press; 123 Seventh Avenue; Suite 164; Brooklyn, New York 11215. All rights reserved. No part of this book may be reproduced without the publisher's written permission, except for brief quotations in reviews.

Printed and bound in the
United States of America

Library of Congress Cataloging-in-Publication Data

Alper, Matthew,
        The "God" Part of the Brain: A Scientific Interpretation of Human Spirituality and God / by Matthew Alper

        4th Edition

                p. cm.
                Includes bibliographical references and index

ISBN 0-9660367-0-0 (pbk.)
I. Title.

Cover art by Phil McGovern

Photograph by Michael Watson

www.godpart.com

ACKNOWLEDGMENTS

I would like to thank my parents, Joan and Jud, and sister, Elizabeth, for their enduring support; Dr. E. Fuller Torrey and Dr. Arthur Rifkin for fixing me; John Stern; Art Bell; Coliseum Books; Dr. Arnold Sadwin; David C. Noelle, Ph.D.; William Wright; Barnes & Noble; Sherry Sanderson at McNaughton & Gunn, Inc.; Dr. Jessica Krant; Joe Fried; Tonya Watson; Lenny Greenberg; Michael Watson; Michael Cunio and Mike Inzirillo; Albert Fernandez; Al Tenzer, and all those innumerable others who have helped me along the way.

Dedicated to the family of God,

...with all my condolences.

"GREAT IS THE TRUTH
AND MIGHTY
ABOVE ALL THINGS"

-THE APOCRYPHA

I Esdras iv, 41

CONTENTS

THE "GOD" PART OF THE BRAIN:

A SCIENTIFIC INTERPRETATION
OF HUMAN SPIRITUALITY AND GOD

– PROLOGUE –

"Man finds himself in the world, or has been thrown into it, and as he stands facing the world he is confronted by it as by a problem which demands to be solved."
— Nicholas Berdyaev

"I want to know God's thoughts...the rest are just details."
— Albert Einstein

Knowledge is power, and it is precisely our species' capacity to reason, to deduce knowledge, that has secured us the title of "The Most Powerful Creature On Earth." Human beings reason because we are compelled to do so. Our survival depends on it, for with every new piece of information we accumulate, whether it be as individuals or a species, we become that much better equipped to master our worlds and therefore to survive.

In addition to this purely practical need to acquire information, our species also seeks knowledge in the hope that it might provide us with a sense of meaning and purpose in our lives. In this regard, our species is unique from all others in that, complementary to our more vital needs, humans possess what we might call "spiritual" needs as well. No less than our bodies crave food, we long to understand our purpose in the universe, our reason for being.

And so, throughout the centuries, our species has gathered information, not just to better master and manipulate the world but also to understand our place in it. We compile information as if each new discovery will contribute yet another piece to some sort of cosmic jigsaw puzzle, which, once complete, we hope will bear us a definitive picture of why we are here.

Every day, under the auspices of science, humankind unravels another of the universe's mysteries, trusting that each new discovery will add yet another piece to this ultimate puzzle. From the innermost particles of matter to the outermost expanses of the cosmos, our ignorance is constantly being replaced with knowledge and understanding.

Yet with all of our knowledge, there still remains that one ever-elusive piece of the puzzle, that one mystery which looms tauntingly over all of the physical sciences, and that is the problem of God. This, more than anything, appears to be humankind's ultimate challenge, that one riddle which - should it ever be resolved - might possibly grant us the definitive picture we've been so painstakingly searching for. Perhaps it is here we will find that underlying the question of God's existence is the answer to man's.

But before we broach the question of God's existence, we must first, as Socrates taught us, define our terms. Exactly who or what are we referring to when we speak of God? Are we speaking of the Greek Gods, Egyptian, Norse, Yoruba, Aztec, of Zarathustra, Buddha, Yahweh, Brahma, Krishna, Amen-Re, Allah? How is it possible to address the question of God's existence, when the word means so many things to so many peoples?

As unique as these various gods might appear, if we take a closer look, they apparently share some very distinct similarities. Perhaps if we were to strip all of these various gods of their more extraneous qualities and were only retain those common to them all, we might establish one entity we could characterize as the "Universal God."

So what might some some of these universal qualities be? What is the Universal God? How shall we define Him? (Though I really mean "It" when I refer to God, for convenience's sake, I will use the masculine form.)

Since the dawn of our species, every single world culture has maintained a dualistic interpretation of reality. In other words, every world culture - no matter how isolated - has perceived reality as consisting of two distinct substances or realms: the material (physical) and the spiritual.

According to this universal perception, objects that belong to the material realm are tangible, corporeal, that which can be empirically experienced or validated i.e., seen, felt, tasted, smelled, or heard.

Objects that exist as a part of this realm are subject to the physical forces of change, birth, death, and decay and are, consequently, perceived as existing in a state of constant flux, temporary, fleeting, ephemeral.

On the other hand, our species equally perceives the existence of a spiritual realm. As this realm transcends the nature of the material universe, things comprised of spirit are immune to the laws of physical nature, that is, to change, death, and decay. That which exists as a part of the spiritual realm is consequently perceived as being fixed, permanent, eternal, everlasting.

Since all cultures perceive their gods as the embodiment of the spiritual realm, we could say that the Universal God represents the essence of all spirit. As spirit is infinite, indestructible, and everlasting, the Universal God, as the essence of spirit, must therefore possess these attributes as well.

Before the Universal God, there was nothing. He is universally perceived as the first cause of all that exists, the self-created creator. The great pageant of matter, from the atoms, planets, and stars, to the multifarious forms of life, all constitute the various ways the Universal God has chosen to manifest Himself. Because the Universal God permeates all that exists, He is both omnipresent and omniscient.

The Universal God represents the embodiment of existence in all its perfection, the supreme and absolute being. As Euripides said, "If God is truly God, he is perfect, lacking nothing." Anything less than this, just the slightest compromise, would necessitate something other than, something inferior to God. There can be no gray area, no in-between. Either God exists as the definitive force in the universe, or He does not exist at all.

But why should I trouble myself with such ethereal matters? Why should the problem of God's existence be of any concern to me? Well, let's suppose for the moment that God does exist. How might this personally affect me?

In accordance with my working definition of a Universal God, if everything that exists does so as one of His extensions then I, too, must exist as such. Now if I exist as an extension of God, and God is spirit, then I, too, must be conceived at least partly in spirit as well; I, too, must possess some measure of the infinite and the eternal within me. In essence, if God exists, then I am immortal, free from the threat of imminent death.

Furthermore, if God exists, my life is replete with meaning. If God exists, then, as the absolute being, his laws must represent absolute truths. It therefore becomes my life's mission to understand these laws, these truths, so that I might live in accordance with them. Moreover, as an extension of God, only by learning to understand Him can I ever really learn to understand my own self. Gaining knowledge and insight into the nature of my creator thus becomes life's intrinsic purpose. With God, I am conceived in meaning.

And if God does not exist? Then I am no longer an extension of some great transcendental force or being, no longer one with any exalted spiritual power or realm, no longer infinite or eternal. In short, I am mortal. And if I am mortal? Then death is final, the decisive end of my existence. If I am mortal, then this one short life, these few fleeting and whimsical years will be the only ones I will ever know. And when they're over, "Out, out, brief candle!" This person *I* call *me,* the sum of my conscious experience will be snuffed out for all eternity. Without God, there is no spirit, no vital transcendental realm or being. Instead, I am abandoned to the spiritless forces of a coldly indifferent and mechanistic universe, an expendable cog in a soul-less machine - here today, gone tomorrow - a random event in an arbitrary universe, no more or less significant than a speck of cosmic dust. In essence, without God, life contains no intrinsic purpose or meaning.

Furthermore, without God, there are no absolutes. All of our laws, our morals, our so-called "eternal truths," are rendered subjective conceptions, man-made devices, as flawed and imperfect as the humans who created them. Good and evil become relative terms, devoid of any true meaning. Without God, there is no absolute moral order in the universe. We become existential orphans, barren of purpose, truth, or soul, forever lost to the vast and meaningless void.

So either God exists, and I'm immortal, or God does not exist, in which case this brief and purposeless stay here on earth is all I will ever know. With God all is saved; without Him, all is lost including hope. Between His existence and non-existence, there is no gray area; there is no in-between. Nothing lies between the finite and the infinite, between the temporal and the eternal, between fleeting meaningless life and purposeful immortality. And so, as man finds himself in this world, or has been thrown into it, and as he stands facing the world, it is the problem of God's existence which demands to be solved before all others.

From the moment this whole disquieting notion first occurred to me, sometime during my mid-teens, those years of which Wordsworth wrote, "bring upon the philosophic mind," I realized that my life's primary pursuit would be — if it were at all possible — to acquire clear and distinct knowledge of God. Does one exist or not? But how could I do otherwise? Was this not literally a matter of life and death, even more so, of eternal life versus eternal death? What should concern me more than my own mortality? If there was any one thing I could say I knew with any certainty in life, it was that one day I was going to die. The question now was, would death mark the decisive end of my existence or the advent of a new beginning?

Here I was, at a time in my life when I was being asked to make such critical decisions as what college did I want to go to, and what would I study once I got there. Only how was I to concentrate on such trivialities with the problem of my own mortality left unanswered? How could I justify an interest in tomorrow while ignoring the greater question of where I would stand against all eternity?

Without answers to such ponderous questions, eternity stood before me like a cosmic brick wall. The universe began to take on the proportions of an unfathomable void, one which, if not sated with knowledge of God's existence, I was beginning to feel would eventually consume me. I needed answers. I needed to know. Was this a world of genuine magic and miracles, or wasn't it? I needed some tangible, reliable, verifiable data that would either prove or disprove God's existence, once and for all.

And so, like an Arthurian knight in search of his Holy Grail, I said goodbye to the conventional world and, instead, rode off alone into the vast dark forest of existence in search of an answer to that ultimate problem: Is there a God? Does one exist? I spent many years lost in those seemingly impenetrable woods, often depressed and despondent, thinking I would one day die there without ever having ascertained a single thing.

But at last, I have returned...furthermore, with what I believe might be the answer.

BOOK I

<u>THEORY'S EVOLUTION</u>

"According to the doctrine of chance, you ought to put yourself to the trouble of searching for the truth; for if you die without worshipping the True Cause, you are lost."

- Pascal

"The unexamined life is not worth living."   - Socrates

ONE

<u>THROWING ROCKS AT GOD</u>

"The Caterpillar and Alice looked at each other in silence for some time: at last the Caterpillar took the hookah out of its mouth, and addressed her in a languid, sleepy voice.

"Who are you?", said the Caterpillar.

Alice replied rather shyly, "I--I hardly know, sir, just at present--at least I knew who I was when I got up this morning, but I think I must have been changed several times since then."

-Lewis Carroll

By the time I was twenty-one, my quest for knowledge of God's existence had taken several unexpected turns. In this time, I had searched the world's myriad religions, only to find myself continually frustrated by a gamut of flaws and inconsistencies in all their logic. I had investigated the various paranormal phenomena, only to encounter a trail of false claims and chicanery. I had experimented with the mind-altering effects of psychedelic drugs as well as transcendental meditation, only to undergo a series of distorted sense-experiences, none of which had brought me any closer to acquiring knowledge of any spiritual reality. As a matter of fact, if anything, they had only served to draw me farther away. This was due to the fact that while exploring the effects of LSD, I had a "bad trip," one that precipitated a severe clinical depression complemented by a dissociative, depersonalization, and anxiety disorder. For a year and a half, I suffered this unfortunate state, until finally, with the aid of pharmacological drugs, I was restored to my original, relatively healthy self.

Though it may have come at a very high price, I managed to garner some extremely valuable information from this otherwise unwanted and wretched experience, information regarding the nature of my allegedly immortal human soul.

According to the various belief systems [religions] I had thus far encountered, the human soul was supposed to be a fixed and permanent agent, unalterable, and everlasting. Again and again, I was told that when I died, though my physical body would perish, "I," the sum of my conscious experience, the essence of my thoughts and feelings, that which was perceived as constituting my spirit, my immortal soul, would endure for all eternity. The fact, however, that my conscious self had been so ravaged, scrambled, and defiled in the past year and a half convinced me that there was no fixed or eternal essence in me.

Twice, in the past year and a half, I had undergone two complete transformations of my so-called "eternal self." First, my conscious self was transformed into something "other-than-it-previously-had-been" by psychedelic drugs. Then, a year and a half later, my original self was restored, this time by an anti-depressant drug known as a mono-amine oxidase inhibitor. But I thought consciousness was supposed to be conceived in spirit, fixed, eternal, immune to the laws of physical nature. If this were true, how was it that the core of my conscious experience had been altered, twice now, by ingesting physical substances? How was it that a combination of molecules - raw matter - could affect something as allegedly ethereal as consciousness, that which was supposed to represent my immutable transcendental soul? To believe that matter could actually alter one's spirit, that it could impact upon one's soul would be the equivalence, it seemed, to believing that one could throw rocks at God. If spirits or souls, truly existed, they should not be able to be affected by matter.

Seeing that my conscious self, my allegedly immortal soul, had been so susceptible to the effects of physical substances convinced me that the soul must be composed of a physical substance as well. A soul that isn't composed of spirit, however, cannot be a soul at all but must rather be the manifestation of some strictly physical phenomenon. Now if consciousness represented the effect of a strictly physical phenomenon, then there would be no place to turn for answers other than the physical sciences.

Up until this point, I always had the greatest respect for the physical/natural sciences. I was always impressed by their ability to make sense of things, their capacity to so effectively reason. Whereas in the past, I had admired the sciences, I now revered them. The sciences had saved my life. I was indebted to them. God didn't save me. I didn't save me. Science, the tool of reason, had saved me. I was my own living proof that science worked. And so, with the same faith that many placed in a god or religion, I would now place mine in science.

Not that I didn't have faith in science before this. Every time, for instance, that I flipped a light switch, one could say I had faith that the lights would go on. The difference now was that, whereas in the past I had taken my faith for granted, I was now a staunch and devout believer.

As I saw it, science had resolved the riddle of the human soul. Science had proven it could come up with chemical formulas that could manipulate the contents of one's sensations, cognitions, emotions, and perceptions in almost whatever way it saw fit. It could electrically or chemically stimulate parts of one's brain in such a way that it could make one either passive or aggressive, tranquil or manic, happy or depressed. In essence, science could manipulate one's thoughts, feelings, and perceptions as if it were pulling the strings of a marionette, making one dance as it so desired.

I was now convinced that the mind, which I previously believed to represent my soul, instead represented the workings of my organ, the brain. There was no soul. There was no ghost in the machine. There was no transcendental component within my head. My thoughts were not the manifestations of some ethereal force or will but rather the consequence of electrical and chemical impulses being registered throughout my brain, impulses that generated a host of sensations, perceptions, emotions, and cognitions in me.

Regarding that question I had been asking myself all these years, of whether human consciousness represented a strictly physical phenomenon or the effects of some transcendental element that resided within my head, as far as I was now concerned, this problem had been resolved. The riddle of the human soul was answered. From hereon, I would interpret the origin of all perception, sensation, emotion, and cognition from a strictly physical/organic/biochemical/mechanistic/rational, that is, a scientific perspective.

As confident as I now was that there was no such thing as a transcendental soul, I still found myself plagued by that more essential question of God's existence. As God represented the creator and embodiment of all things spiritual, not until I could resolve the problem of His existence could I be absolutely certain that there was no such thing as a soul. Until I possessed some rational explanation through which I could interpret the nature of God, there was still a chance there existed a spiritual reality. And as long as there might exist a spiritual reality, there remained the possibility that souls existed as well. Before I was ready to commit to anything, I needed to resolve the more all-encompassing problem of God.

As the physical sciences had helped me to rationally interpret the underlying nature of consciousness, I now wondered if it would be possible to apply this same tool of reason to resolve that ever-persistent problem of God's existence. Could the physical sciences crack that nut as well? Up until now it hadn't come close. From biologists to astro- and quantum physicists, no one had ever advanced anything resembling a scientific interpretation of God. But why was this? Did God truly exist, only He was beyond our grasp, beyond the range of human comprehension? Or was there a genuine and tangible solution, simply no one had discovered it yet?

As a now staunch believer in the ways of science, I felt there must exist a rational explanation for everything. As a scientific idealist, I found myself inclined to believe that nothing was beyond our reach. If it could be dreamt of, it could be reasoned through.

My course was now defined. I would be a scientist. I would accumulate all the scientific knowledge I possibly could, and then, once this was accomplished, once I had familiarized myself with all the various scientific disciplines, only then would I justifiably recommence with my quest for knowledge of God.

But wait! What if it should turn out that science was just another form of psychological indoctrination, a new religion for a new world? What if science was no more founded in truth than any of the other self-glorified creeds I had thus far encountered? Perhaps science was just another mythology, one which sought to replace the antiquated gods of the past with newer ones, a new belief system for the twentieth century. Perhaps scientists were just the high priests of a new faith, one that, instead of referring to gods, referred to particles that were just as incomprehensible and elusive. Perhaps science was just another false paradigm. Then again, perhaps it was not. Perhaps science was a genuine tool by which human beings could gain a clearer and more distinct insight into the underlying nature of reality. So which was I to believe? How could I prove that scientific facts were any more reliable than religious ones? It was time to define my terms, time to investigate the investigator. Before I would blindly place my trust in the scientific process, before I submitted myself to a lifelong quest for "scientific" knowledge of God, I would investigate the nature of my newfound faith. "What," I had to ask, "exactly is science?"

TWO

<u>WHAT IS SCIENCE?</u>

"Science is the attempt to make the chaotic diversity of our sense-experience correspond to a logically uniform system of thought."   -Einstein

"There is no such thing as absolute certainty, but there is assurance sufficient for human life."   -John Stuart Mill

In order to justify my quest, I conducted an investigation into the nature of the physical sciences, of science itself. This is what I found: What is science? Since this is a rather large question, I will do my best to explain it in the most conceptual terms I am able.

Before I begin, let me state that no matter how much one may place his faith in science, he must realize that at no time can it ever represent anything more than just another belief system, just another way by which humans can choose to interpret reality. I say this not out of any lack of conviction but only because not even science can guarantee anything with absolute certainty. Nothing can! Who, for instance, could guarantee with one hundred percent certainty that his or her life is anything more than just an illusion or a dream? As written by a sage over two thousand years ago, "Once upon a time, I, Chang-Tzu, dreamed I was a butterfly, fluttering hither and thither...Suddenly, I was awakened...Now I do not know whether I was a man dreaming I was a butterfly, or whether I am a butterfly now dreaming that I am a man." (MacKenzie, 1965).

Nothing is certain! No wonder one of the wisest men to walk the earth [Socrates] lived by the principle that all he knew was that he knew nothing at all.

Nevertheless, with that necessary qualifier aside, let's presume for the moment that this experience we call life isn't a dream. Let's suppose for the moment that we truly do exist as what we imagine and that our experiences are, more or less, "real." Even so, it would still be impossible for us to ever possess absolute knowledge of anything. Let me explain...

The only means we, as human beings, have to interpret reality is through information acquired through our physical sense organs. Through our eyes, we absorb photons of light; we *see* the world. Through our ears, we absorb vibrations; we *hear* it. Through the nerve endings that cover the surfaces of our skin, we experience differences in pressure and temperature; we *feel* the world. Through our noses and tongues, we absorb chemicals; we *smell* and *taste* it. Before we acquire knowledge of our world, all information must first pass through these physical sense organs. Our sense organs therefore determine the manner in which we perceive reality. As each species possesses its own unique set of sense organs, each must experience and, consequently, interpret reality from its own unique and relative perspective.

Common house flies, for instance, have a different mechanism from ours by which they absorb light, that is, they possess a different set of organs we call eyes. As flies sense the world differently from us, they must consequently interpret it differently as well. Just as a fly sees the world from its own unique fly perspective, we see the world from our own unique human perspective. Whereas flies possess fly knowledge, humans possess human knowledge. And just as a fly can only possess fly knowledge and no other, a human can only possess human knowledge and no other. We must therefore accept that our interpretation of reality is no "better" or more "real" than a fly's. It is simply different.

Moreover, it's not just the manner in which our physical sense organs absorb information that frames our perspectives but, just as significantly, the manner in which we then process that same information. For example, when we eat an apple, we feel its texture, smell its aroma, taste its flavor. Not until we integrate these various sense-impressions, are our sensations transformed into a coherent perception of the apple as a whole. Without some kind of internal processor to coordinate the medley of sense-impressions we receive, it would be impossible for us to comprehend our experiences and therefore to survive.

In the least sophisticated organisms such internal processors constitute a single neural pathway. For those animals that possess a slightly more complex processing mechanism, we call this a ganglion. An even more sophisticated processing mechanism, we call a brain. Our brain, our processor, represents the most sophisticated of all. Because each organism possesses its own unique processing mechanism, its own unique central nervous system or brain, each organism must, consequently, interpret reality from its own unique and relative perspective.

Furthermore, it's not just the different species that perceive and interpret reality from their own unique perspectives, but also each individual within each species. Among our own species, each individual possesses his own unique combination of sense organs, that is, his own unique combination of ears, eyes, nose, and mouth. In other words, no two humans have the exact same set of sense receptors. For example, because the physical nature of my eyes is slightly different from my neighbor's, I might perceive the color red a shade lighter than he does. In an even more extreme example, a person with damaged cone receptors might be totally color blind, and will consequently experience what I perceive to be bright red as toneless or gray. Because each individual perceives the world from his or her own unique perspective, each of us must possess our own unique interpretation of reality.

Just as each individual's set of sense organs vary to some degree, so does each individual's brain or processor. Just as no two people possess the exact same eyes, no two people possess the exact same brain. Therefore not only does each individual acquire sensual data differently, but each of us also processes that same data in his or her own unique way.

In addition to these two factors, we must also take into consideration the fact that each individual lives a unique set of life experiences. As this, too, will affect one's cognitive development, it will also affect the manner in which one will interpret reality.

There are therefore three variables that determine the way each species, as well as each individual within each species, interprets reality. These include the physical nature of an organism's sense organs, the physical nature of its processor (or brain), compounded by the content of its life experiences.

With these three variables in mind, let's imagine that two amoebae, two houseflies, two chimpanzees, and two humans are all perceiving the same sunrise. As each of these individual entities absorbs and then processes the sun's radiated light in its own unique way, who could possibly say which of their experiences is the most genuine or "real?" What organism could dare claim that it sees the "real" sunrise? Which organism could say that its experience of the rising sun's red color is any more authentic? Red is a man-made concept that has no bearing on the actual physical universe. Though we may interpret the sunrise as being red, the sunrise "in-itself" is not. This is just the name we attribute to the manner in which the mean of our species interprets a particular wavelength (six hundred nanometers) of light as it falls upon our retinas. In essence, our perceptions bear us relative information regarding our world, as opposed to anything absolute.

As each of us perceives the world from our own unique and therefore relative perspective, all knowledge must consequently be relative as well. In the words of the German philosopher Immanuel Kant, we can never know a "thing-in-itself" but only a "thing-as-we-perceive-it." Consequently, we can never know anything with absolute certainty. If this is true, one might rightfully ask: why seek to know anything at all?

The answer to this is simple. Regardless of how relative our perspectives might be, we, nevertheless, possess the capacity to perceive a close or common *enough* approximation of things to provide us with practical information regarding our world. This is why, for instance, if we were to take a roomful of people all looking at the same rock, and we were to ask them what they saw, though each individual might offer his or her own unique interpretation, each will generally agree that the object at hand is indeed a rock. If, among this same roomful of people, some claimed to see a dog, others a shoe, while others a banana, then perhaps we'd be in for some trouble. Fortunately, for our species, this is not the case. Fortunately, our sense organs are consistent *enough* that if we were to place an object such as a rock in front of a roomful of people, all will generally agree that it is a rock they are perceiving. Though we may never know a "thing-in-itself," that is, though we may never possess absolute knowledge of anything, our perceptual organs and internal processing mechanisms provide us with a consistent *enough* account of the world to provide us with practical and reliable data. As a matter of fact, our perceptual organs have yielded so much practical and reliable data that we have been able to develop entire scientific disciplines from them. These disciplines have helped us to cultivate such technologies as the electric light, microwave oven, nuclear energy, artificial organs, space ships, antibiotics, electron microscopes, and computers, to name a few.

So what is science's secret? How does it allow us to take our perceptions of things and transform them into an electric light or a microwave oven? What application of knowledge is this that it has furnished us with such a vast wealth of life-enriching technologies? Simply speaking: how does science work?

Science relies on a very strict process known as the scientific method, a process whose principles were originally outlined by Rene Descartes in his book, *Discourse on the Method of Properly Conducting One's Reason and of Seeking the Truth in the Sciences*. In this work, Descartes suggested that in order to procure "clear" and "distinct" knowledge of things, one has to apply a strict set of guidelines to the manner in which one conducts his observations. Descartes referred to these guidelines as the scientific method. And what is this scientific method? Without providing a detailed explanation of Descartes' own principles, I will attempt to offer a more conceptual interpretation.

The scientific process operates in two phases, the empirical and the statistical. In the first phase, a scientist seeks uniform patterns in the universe based on empirical observations (empirical meaning based on data received by our senses, as this is, after all, the only means we have). For example, an early human happens to notice, based on information received through his sense organs, his eyes, that the sun rises in the east. The next morning, he notices the same thing occur. After several more observations, this nascent scientist begins to recognize a pattern. Based on his initial observations, he may surmise that perhaps the sun, as a rule, rises from the east.

Based on his observations, this person will now propose a theory. He will hypothesize that the sun, as a rule, rises from the east. Since he has yet to confirm his theory, his assertions are, for the time being, purely hypothetical. After all, a few simple observations are hardly any basis for placing unconditional faith in something.

It is now, in the second phase of the scientific method, that a scientist must perform a series of tests that will either verify or refute his original hypothesis. He may, for example, decide to observe the sunrise for several more years, allowing each morning's observation to represent one more piece of evidence to confirm his hypothesis. This is where the statistical phase enters the picture.

After our scientist feels confident that he has obtained *sufficient* statistical evidence to support his theory (that the sun rises from the

east), he will disclose his findings to those around him, more specifically to the rest of the world's scientific community. Now it is the
duty of the scientific community to review this person's hypothesis
by performing their own series of tests. This is necessary as the conclusions of one sole observer should never be accepted as *adequate*
proof of anything. What if, for instance, our original scientist was
making up the results just to get his name in the papers, or perhaps he
was simply too ignorant to know the difference between east and
west?

It is at this point that other scientists will perform their own tests
meant to either confirm or invalidate the original scientist's hypothesis. Perhaps some of these scientists will duplicate the original scientist's experiments to see if they, too, will get the same results. Others,
meanwhile, may devise whole new means of testing the theory. One,
for instance, may wish to see whether or not he will obtain the same
data from some other part of the planet. Perhaps in Africa or Asia the
sun rises from the west.

As this process continues, one by one, our ever-skeptical scientific community will conduct as many tests as it can think of before
offering to support a theory. Only after a *sufficient* amount of supporting statistical data is obtained might the scientific community be
willing to give credence to a theory, in this case, that the sun does
indeed rise in the east.

Keep in mind, statistics still do not reflect certainties. Though the
sun may have consistently risen in the east for as long as humankind
has recorded this phenomenon, the assumption that the sun rises from
the east is still just a theory. Just because the sun has risen in the east
every day up until the present doesn't necessarily mean that it will do
the same tomorrow. How, for instance, can we know with absolute
certainty that the sun won't explode this evening for reasons beyond
our knowledge? We don't. What we do know is that the sun has been
rising in the east for so long and with such consistency that it *most
probably* will do the same thing tomorrow – not certainly, just most
probably. Even Einstein recognized that though no one single experiment can ever prove a theory absolutely right, all it takes is one to
prove a theory wrong. (For example, should the sun rise in the west,
just once, there goes the entire theory.) Scientists do not, therefore,
claim to be able to "see" into the future but only to predict, within a
certain degree of accuracy, based on probabilities, what may or may

not occur.

But if science is based on mere probabilities (as opposed to certainties), why should we place so much faith in it? Why practice science with such conviction? The reason is that although the whole of science may be based on probabilities, it still represents the most accurate and reliable source of information any system has offered us so far. Though our local meteorologist may sometimes provide us with an inaccurate forecast, how often do we choose to turn to our local priest, shaman, or tarot card reader for tomorrow's weather? Though scientific method may be based on mere probabilities and therefore imperfect, it has proven itself, time and time again, to represent the most reliable and accurate source of information we have.

Once the scientist has *probable* cause to give credence to a theory, once he has faith that the pattern he has recognized occurs with a *sufficient* degree of consistency, he will then use this newfound information to elicit even more. One deduced "fact" can be used to deduce the next. Once our scientist accepts that the sun rises in the east, he is now armed with yet one more fact with which to examine the world, one more piece of the puzzle with which to try to grasp the greater picture. In his search for answers, the scientist will utilize his findings to uncover even more elusive patterns. In this way, science is constantly building upon itself.

One of the fundamental principles of science is that every action has an effect, which in turn means that every effect has its cause. Once a theory has been verified, a scientist might want to know *why* this occurred. Once, for instance, the scientist accepts that the sun rises in the east, he may want to dig deeper into the mystery of this phenomenon by asking: why does it rise this way? Is it because a sun god is pulling it up from the east by a magical string, or maybe because the sun revolves around the fixed Earth from that direction? Presuming that the sun rises in the east, the scientist may now search for yet an even deeper understanding of why this phenomenon occurs.

With the assistance of various tools that can be used to enhance our empirical powers of observation (such as a telescope with which to augment our vision), a scientist can dig perpetually deeper into the mysteries of the physical universe, acquiring information, one piece at a time, until he has acquired as much knowledge as is humanly possible.

Now there are those who resist science, those who deny its capacity to reliably interpret reality, those who consider it a sham, an artifice, a means of deceit. They refer to science as the devil's plaything, a conspiracy developed to contradict one's religious beliefs. Take, for instance, those who support the Judeo-Christian interpretation of the Earth's creation. Such people reject Man's evolution from the primates. They refute the idea that the Earth (as well as life) is a few billion years old. Regardless of how much their beliefs (e.g., that the world was created in seven days, approximately six thousand years ago) may contradict libraries full of carefully documented scientific data (data acquired through the exact same methodology that gave us the electric light and nuclear energy), they insist that their viewpoint is correct. How is it that such people can refute such well established data and yet, in the same breath, turn on their electric fans when they are overheated or take antibiotics when they are ill? How can such people spurn the sciences one day and then gladly partake of its fruits the next? How do they justify their acceptance of such medical technologies as gene therapy or cloning while, at the same time, continuing to deny the same evolutionary principles from which these advances are founded? There is no compromise. One must either accept the doctrines of science – OF REASON – or one must reject its principles altogether. We either trust in the scientific method or we do not.

One problem many religions have with science is that it represents a source of constant contradiction. For example, in the old days, if the land was dry, men prayed for rain. Since they didn't understand the underlying physical cause of this phenomenon, they believed that the rain's fall was determined by the impulses of those who lived beyond the clouds, by the wills of the gods. How else were humans to explain such a thing? They couldn't. It took humankind thousands of years of scientific discovery and research before we understood the nature of evaporation and condensation of water molecules, that is, of rain. But we needed some sort of an explanation. What else were we to do? Accept that it rained for absolutely no reason whatsoever? We are compelled to understand. It is human nature to pursue the underlying nature of things.

Today we know better than to believe that rain is produced by the whims of gods. Today, we know that rain occurs because of a series of physical causes and effects. In this way, science has emasculated the old gods. It has stripped them of their powers and has instead allotted them to a source that is wholly neutral, one that is indifferent to the affairs of men, one scientists refer to as "the forces of nature."

Now I can certainly understand why humans would desire to believe in a god, in a force that cares about us, that treats us as its favored creatures. Believing in a god provides us with a sense of purpose, with a sense that we are immortal. But should we believe in such things if it's at the expense of everything that corresponds with scientific method, with reason?

And so, at the age of twenty-one, I decided to place my faith in the natural/physical sciences. And why not? At this point, I had every reason to believe in the logic of the physical universe and none, whatsoever, to believe in any spiritual reality. Until proven otherwise, I would pursue all things, including the nature of God's existence from a strictly physical, that is, scientific standpoint.

Only how was one to use science to find God? Into what constellation does one point his telescope? What slide is one to place under the microscope?

... And so, my quest continued.

THREE

A VERY BRIEF HISTORY OF TIME
or
EVERYTHING YOU EVER WANTED TO KNOW ABOUT THE UNIVERSE BUT WERE AFRAID TO ASK

"The aim of all intellectual pursuits including science, philosophy, and art, is to seek unity in the midst of diversity, order in the midst of complexity. Their ultimate task is to fit multifarious elements into some kind of compact, cohesive, apprehensible scheme."   — D.E. Berlyne

"And I gave my heart to seek and search out by wisdom all things that are done under heaven."   —Ecclesiastes

"Canst thou by searching find out God?"
                                        — Bible, Job 11:7

So off I went, full speed ahead, searching through numerous scientific tomes...for God. There was anthropology, physics, biology, psychology, physiology, geology, astronomy, chemistry, cosmology, to name a few, each one a school unto itself.

The more I studied the various sciences, the more I realized how much they were all interconnected. It was as if the scientists had somehow made the mistake of breaking this one unified history of the entire physical universe into several separate epochs or categories, not realizing they were each integrally linked to one another in the most essential way. Science, I began to realize, was simply the study of the history of the entire physical universe from the dawn of time.

As I embarked on my newfound quest for a scientific interpretation of God, I decided to begin with physics, as it seemed to address nature's most fundamental principles. From physics, I learned how the universe was born, approximately fifteen billion years ago, at which time all the matter in the universe was condensed into one single, solitary point of pure energy. The pressure within this point was so great that it erupted in an enormous explosion, which, in turn, released all of the universe's energy outward into vast space, an event scientists refer to as the "Big Bang."

As Einstein taught us, energy and mass (matter) are interchangeable: $E=MC^2$ – Energy equals mass (matter) times the speed of light (approximately 186,000 miles per second) squared (multiplied by itself). What this essentially means is that if mass (matter) is accelerated to a high enough speed it will become energy. Inversely, should energy be slowed down, it will settle into matter. And so, within one-millionth of a second after the universe's initial eruption, energy began to settle into its first material particles. By one ten-thousandth of a second, forces inherent within these first infinitesimal particles prompted them to bond with one another to form larger infinitesimal particles. Three minutes later, after these first "sub-atomic" particles had bonded with one another, they settled into the first stable form of matter known as "atoms," lithium, deuterium, and hydrogen atoms to be exact.

For the first several billion years after this initial eruption occurred, hydrogen represented the predominant form of matter to exist within the universe. At this time, the universe existed as a swelling expanse of gases consisting chiefly of hydrogen atoms, which, due to the initial force of the "Big Bang," was being propelled further outward into vast space.

The law of gravity states that all matter is attracted to all other matter. It was this force inherent within all of these hydrogen atoms that prompted them to gravitate toward one another, causing them to congregate into ever-enlarging gaseous clouds.

Now there were two forces working on the hydrogen atoms simultaneously, one that propelled them outward into space and another causing them to gravitate toward one another. This second force continued to act on the atoms until they had gathered into humongous clouds consisting of huge masses of hydrogen. Because the force of gravity always draws matter towards an object's center, all of the hydrogen atoms within these enormous clouds collapsed upon the cloud's cores. The weight of all this hydrogen collapsing upon itself created a tremendous amount of pressure within these clouds' cores. When the pressure within the cores became more than the individual hydrogen atoms could withstand, they began to fuse.

When this occurred, four hydrogen atoms were compressed or "fused" together to form a heavier atom we call helium, the newest stable form of matter or "element" to exist within the universe. When four hydrogens fuse to create helium, not all of their mass is retained within the helium. Instead, some of the hydrogen's mass is lost as energy, radiated outward in the form of heat and light. The moment one of these hydrogen clouds begins this fusion process, we refer to it as a star. Our own sun represents a perfect example.

Millions of years after a typical star is born, after it has fused much of its hydrogen atoms, it begins fusing its heavier element, its helium. When helium atoms fuse, they are transmuted into the even heavier element of carbon. As this process continues, newer, heavier atoms or elements are created within a star's core. After a star depletes itself of most of its fusible matter, it becomes unstable, often causing it to erupt in a tremendous explosion called a supernova. Such supernovas cause such newfound elements to be dispersed throughout the ever-expanding universe.

It was at around this point I noticed that my physics texts were coming to a close and my chemistry books just beginning. It seemed that once these newly created elements began interacting with one another, for some arbitrary reason, the history of the universe had been divided into a whole new field of study. It was as if someone had decided to break the history of the universe into chapters. In finishing "Physics," I had just completed chapter one in this cosmic serial. It was now time to move on to the next installment in the history of the Universe – Chapter Two: Chemistry.

Physics had outlined the essential forces of nature, forces inherent in all matter. When dealing with how these forces affected matter's smallest particles, this was referred to as particle or atomic physics. When dealing with how these forces affected the interaction of much larger objects such as planets or stars, it was called astronomy. When dealing with the full scope of all the energy and matter that existed within the entire physical universe, this was called cosmology.

After physics had left me with an interpretation of the various atomic forces, as well as how the diversity of elements were formed, physical chemistry sought to explain the dynamic involved in those interactions that took place between the varying atoms or elements. Since each new element created within these burning stars consisted of a different number of electrons (a sub-atomic particle carrying a negative charge), each atom carried a slightly different electrical charge from all others. Based on their relative charges, some of the differing atoms began to bond with one another to form more stable particles known as compounds or molecules. Chemistry sought to interpret the unique set of properties that each one of these new atomic combinations contained as well as how they reacted with one another. One sodium atom and one chlorine atom, for instance, have a propensity to bond with one another, creating a compound we call sodium chloride, more commonly known as salt. With this new diversity of atoms being dispersed throughout the universe, an abundance of new molecular combinations began to emerge. From its humble beginnings, when the universe consisted of just one single element, it had evolved into a complex array of physical compositions.

Depending on such variables as pressure or temperature, any compound could exist in one of three forms – as a solid, a liquid, or a gas. Many of the compounds, as they existed in solid form, were referred to as minerals. Due to the effects of electromagnetic as well as gravitational forces, these minerals began to gather into ever-larger formations.

Quick cut to astronomy: Nearly five billion years ago, about ten billion years after the universe was originally born, our sun was formed from a tremendous rotating cloud of gases. Although most of this particular cloud consisted of hydrogen, as many stars had already been born and died, it also contained a mix of other elements as well. As the core of this rotating ball of gases consolidated to become a star, some of the heavier elements dispersed around the cloud's periphery began to amalgamate into large mineral clusters made up of a mix of the star's heavier elements.

When one of these large mineral cluster flies too close to a star, it is drawn in by the star's enormous gravitational pull and absorbed into that star. If the cluster's momentum exceeds the star's gravitational pull, it will spin off into deep space. In the rare case that the cluster's momentum happens to be at equilibrium with the star's gravitational pull, it will become caught in a perpetual elliptical path around the star. We refer to this path as an orbit. When a large enough mineral cluster falls into a star's orbit, we call it a planet. We live on Earth, the third planet from our star, the Sun.

Sometimes smaller mineral formations formed around the planets and fell into their orbits. A mineral cluster that orbits a planet, we call a moon. A star combined with the planets that orbit it, we call a solar system. Our solar system consists of a star [the Sun] with nine planets [Mercury, Venus, Earth, Mars, Jupiter, Saturn, Uranus, Neptune, and Pluto] orbiting around it. On an even larger scope, a cluster of solar systems is called a galaxy. All the galaxies in vast space make up the universe. Meanwhile, back to our star's spinning satellite, back to our planet, Earth.

Enter that branch of science known as geology: Approximately 4.6 billion years ago, the Earth, one of the nine planets in our solar system, third closest to the Sun, was formed. At the time, the Earth was no more than a big ball of molten rock. Not yet possessing an atmosphere to shield it from falling objects, the Earth was constantly being bombarded by stray mineral clusters known as meteorites, which fell into its gravitational field. As these falling objects continued to rain upon the planet, the Earth continued to grow in mass and size.

When these meteorites hit the Earth, tremendous explosions occurred that unleashed tremendous amounts of heat energy. This caused the meteorites, as well as those portions of the Earth they had hit, to melt. Many of these meteorites carried trapped gases within them. Once they hit the Earth and became molten, these trapped gases were, consequently, released. Since gases are light and volatile, they have a tendency to fly off into space. Some planets, such as Mercury, are so small that they don't possess a strong enough gravitational pull to retain such volatile particles. Some planets, such as Jupiter, on the other hand, are so large that their gravitational pull causes its gaseous elements to be so firmly drawn to the planet's surface they become condensed into liquid pools.

The Earth, however, was neither too small to retain its gaseous elements, nor was it so large that it compressed them to its surface. It was neither too close to the Sun (the heat of which affects the volatility of these gases) that they were propelled into space, nor was it so far from the Sun that these gaseous elements were frozen into solid form. Instead, the conditions on Earth were such that released gases were held within close enough proximity to the surface that they came to form a gaseous shell around the planet. This shell we call the atmosphere. Now when a meteorite got caught in the Earth's gravitational pull, the friction incurred by its rubbing against the atmosphere's shield of gaseous particles caused these meteorites to burn up before they would reach the Earth's surface. No longer vulnerable to the heat-emitting collisions previously generated by the force of these meteorites, the Earth began to cool.

Two of the gases that were often trapped within these falling meteorites were hydrogen and oxygen. As enormous amounts of these two elements filled the Earth's atmosphere, they began to bond with one another to form a molecule commonly known as water. As water vapor began to accumulate in the atmosphere, it became so dense that it eventually succumbed to the planet's gravitational pull, causing it to fall back down upon the Earth's surface in the form of droplets we call rain. When these first waters fell to the Earth, they caused the planet's molten surface to cool even further, in turn, prompting more of the Earth's trapped gases to be released in the form of steam. More water vapor yielded even more rain, which caused the planet to cool even further.

This process continued for nearly a billion years after which two-thirds of the Earth had become covered in water with the other third made up of a hardened mineral shell. Within these oceans of water, there stirred a brew of ammonia, methane, water, sulfur dioxide, and hydrogen.

In 1953 a graduate student, Stanley Miller, conducted an experiment:

> "Miller set up an airtight apparatus in which the four [original primordial] gases could be circulated past electrical discharges from tungsten electrodes [patterned after the primordial Earth's lightning storms]. He kept the gases circulating continuously in this way for one week, and then analyzed the contents of his apparatus. He found that an amazing number and variety of organic compounds had been synthesized. Among these were some of the biologically most important amino acids as well as such substances as urea, hydrogen cyanide, acetic and lactic acid."[1]

Within the confines of a laboratory, Miller had simulated the Earth's chemical evolution. He had synthesized the building blocks of all organic matter, the essence of all life. In doing this, Miller had accomplished what was formerly believed to be the exclusive right of the gods. And yet, here it was, organic evolution without God - just Stanley Miller and his airtight vessel of chemicals, a flame, and a little electricity.

Starting with a composition consisting almost exclusively of hydrogen, the universe had evolved, almost twelve billion years after its birth, to a point where it contained complex chains of macromolecules. Macromolecules that contained carbon possessed such unique properties that my chemistry books had suddenly diverged into a whole new science called organic or bio-chemistry. I now had to purchase a whole new set of texts that dealt exclusively with these complex carbon-based compounds.

Back to the planet Earth: For the next billion years, these complex organic [carbon-based] compounds brewed and churned within the Earth's primordial seas. Within these seas, trillions upon trillions of various molecular combinations emerged, each one possessing its own set of physical and chemical properties. Many of these molecular combinations to emerge were so complex that inherent instabilities caused them to eventually disintegrate back into their contingent parts.

As these complex molecular forms continued to brew in the Earth's seas, new combinations were constantly being formed, each one slightly different from the next. Among these new complex "organic" molecules, some of the variations to emerge possessed the capacity to absorb the Earth's and the Sun's radiated heat and light energies. With this ability, these otherwise unstable molecules could now use these external energy sources to maintain stability.

Even with this newfound capacity, none of these energy-absorbing macromolecules were efficient enough to overcome their inherent instabilities altogether. Being able to use the Sun's energy merely allowed these complex molecular chains to maintain their structural identities for a slightly longer duration than they could before. Even so, it was always just a matter of time before these molecules eventually succumbed to their inherent instabilities and eventually disintegrated back into their contingent parts.

As newer variations of these complex, energy-absorbing, organic macromolecules continued to emerge within the primordial seas, some eventually developed the capacity to produce duplicates of themselves before they disintegrated. These new molecules could now ensure the preservation of their physical identities through the continued existence of their duplicates. Due to the disruptive effects that the Sun's ultraviolet rays had on these molecules, not all of the copies turned out to be identical to the "parent" from which they came. Among these slight variations, most were harmful and worked against the "daughter" molecules' continued existence. Nevertheless, some of these variations happened to be even more energy efficient than their parent molecules. In such cases, the new design would now supersede the old one. As this process continued, more energy efficient molecular combinations continued to emerge.

In time, these self-duplicating macromolecules evolved newer and more developed functions that allowed them to maximize their capacity to maintain stability. Some of the functions these advanced macromolecules had evolved included ingestion (the capacity to absorb energy), digestion (the capacity to assimilate ingested energy), excretion (the ability of the macromolecule to rid itself of any of its digested energy's harmful by-products), locomotion (the capacity to move from one place or position to another) and reproduction (the macromolecule's capacity to form a duplicate of itself). As these self-duplicating, energy-absorbing macromolecules continued to evolve, I began to notice that my organic chemistry books were now evolving into a whole new science called biology.

As with all the other sciences, biology came with its own terminology. In biology, for example, molecules that could perform the aforementioned functions were now called "cells." With the addition of these newfound characteristics, matter previously referred to as organic was now called "living." When a cell made a copy of itself, this was now referred to as "birth." When, in time, an organic macromolecule eventually disintegrated, this was now called "death."

The first forms of life to exist reproduced asexually, meaning they required only one parent cell that would divide into two distinct daughter cells. Once again, due to the disruptive effects of the Sun's radiation, many of these offspring had mutated and consequently represented a slight variation of its predecessor's design. Variations that were more energy efficient were more likely to endure, or, as my biology books now referred to it, to survive. Those most likely to survive were most likely to duplicate themselves and, therefore, to pass on their advantageous attributes or what my biology books referred to as "traits." On the other hand, those variations that were least energy efficient were most likely to be discontinued. My biology books had a very specific term for this organic weeding process. They called it "natural selection." As a result of this process of natural selection, organic matter or "life" continued to evolve.

The first varieties of life to emerge on Earth diverged into two distinct groups. One used the Earth's oxygen to supplement its fuel supply, while the other used carbon dioxide. In order to keep track of these constantly diverging "living" material compositions, biologists categorized them into groups based on their inherent characteristics. Biologists split these first two groups to emerge into two separate classifications known as kingdoms. Those forms that used carbon dioxide to supplement their fuel supply were classified as belonging to the "plant" kingdom, while those which used oxygen were categorized as belonging to the "animal" kingdom. As time passed, these two distinct kingdoms continued to diversify, each producing a vast array of different forms, or what biologists referred to as "species." Within the next three billion years, a plethora of these various species propagated across the planet, blanketing the earth's surface with a thin organic shell.

Three billion years after life had first evolved, the seas were suffused with a variety of these plant and animal forms. It was at about that time that one of these sea-dwelling animals evolved a spinal cord, a protective sheath that surrounded the organism's nervous system and helped to distribute the animal's nerves throughout the length of its body. This represented the beginning of a new classification of animals biologists referred to as the subphylum, vertebrate. As the vertebrates continued to diverge, biologists placed them into separate groupings known as "classes." The first class of vertebrates to emerge were "fish."

About a hundred million years later, some of these fish evolved the capacity to survive on land as well as in the water from which they came. These, biologists classified as "amphibians." About a hundred million years after that, an even newer class of vertebrates had evolved from the amphibians, one which lived exclusively on land. These were the "reptiles."

Within the next fifty million years, some of the reptiles evolved in such a way that their scales were replaced by feathers, their bones became hollow, and they developed the capacity for flight. This class was referred to as "birds." Approximately another forty million years after that, another land-dwelling animal emerged from the class of reptiles. These were the "mammals." Mammals were different from their ancestors, the reptiles, in that their embryos developed from within the mother's body rather than from within an externally incubated egg. Mammals produced milk with which they could feed their young. They were coated with hair, homeothermic [warm blooded], and, most significantly, developed a much larger brain, allowing them to respond to their world in a much more versatile manner than all the Earth's other life forms.

Among the mammals, sixteen subclasses known as "orders" emerged. Examples of some of these orders were rodentia (rats, mice, squirrels, etc.), carnivores (cats, dogs, bears, etc.), cetaceans (dolphins, whales, porpoises), and artiodactyla (cattle, sheep, goats, deer, etc.). About a hundred million years after the mammals first evolved, approximately fifty million years ago, a particular mammalian order emerged known as "primates." Primates differed from the other mammals in that they evolved such adaptive features as binocular stereoscopic vision, enhanced mobility of the digits (fingers) joined by an opposable thumb, larger brains – particularly a larger cerebral cortex (that portion of the brain where memories are stored and most cognitive processing takes place).

As time went on, these primates continued to change until they had evolved into a family called "hominids." Hominids stood upright, as compared to their ancestors which walked on all fours. With the advent of this new adaptation, these animals now had two free limbs with which they could hold, carry, and manipulate objects at the same time they could transport themselves. The hominids continued to evolve until about a hundred thousand years ago when they reached their apex with the emergence of a new species known as "Homo sapiens," more commonly known as human beings. These Homo sapiens evolved vocal cords with which they could enunciate a variety of sounds, thus enhancing their capacity to communicate with one another. Furthermore, Homo sapiens evolved certain structures within their brains that allowed them to organize these sounds in such a way that they could create and speak words – combinations of sounds that represented objects. These words enabled the animal to communicate ideas with advanced precision. These qualities combined with their enhanced capacity to store and process information, made Homo sapiens Earth's most powerful creature.

Before I delve any further into the subsequent studies concerning the nature of this new human animal, I would like to clarify a few things. In a matter of pages, I have jumped from the origin of the first organic matter to the emergence of humankind. But by what process does such an evolution take place? How is it possible that within three and a half billion years a simple cell membrane could have turned into flesh, a vacuole into a complex digestive system, a cellular nucleus into a brain? How could a reptile's scales become feathers or its legs become wings? What kind of organic alchemy or molecular witchcraft was this that could transform creatures from one thing into another? How does such a process take place? To offer an illustration, let's take the example of a human being.

Two cells, a sperm and an egg, meet. These two cells happen to be distinct from all others within the human body, as each one carries only half of its host's chromosomes. Within the sperm cell's nucleus lie half of the father's chromosomes, within the egg cell, half of the mother's. When these two chromosomally incomplete cells meet, that is, when the egg becomes fertilized, the two sets of chromosomes merge and recombine to form one novel and chromosomally complete cell.

This now complete set of chromosomes within the newly fertilized cell is like a blueprint that contains all the necessary information to create a fully developed human. The chromosomes themselves are composed of sections called genes. Each gene contains information to create one or more of what will soon unfold to become that individual's physical traits. For instance, whereas one gene might carry information that will determine a person's eye color, another might carry information that will determine skin color, another that person's sex, height, hair color, etc. This list of physical features goes on until one's entire anatomy - from the shape of one's head to the soles of their feet - has all been accounted for. All this information is stored within the contents of one's genes.

But what are genes? According to the biologist, William Keeton, a gene is a "unit of inheritance; a portion of a DNA (Deoxyribose nucleic acid) molecule."[2] Here is Keeton's technical description of this molecule:

> "The molecule has a ladderlike structure, with the two
> uprights composed of alternating sugar and phosphate

groups and the cross rungs composed of paired nitrogenous bases. Each cross rung has one purine base (any one of several double-ringed nitrogenous bases) and one pyrimidine (any one of several single-ringed nitrogenous bases). When the purine is guanine, then the pyrimidine with which it is paired is always cytosine; when the purine is adenine then the pyrimidine is thymine. Adenine and thymine are linked by two hydrogen bonds, guanine and cytosine by three."[3]

So, genes are made of DNA, a macromolecule consisting of a combination of sugar molecules, phosphate molecules, and nitrogen-based molecules, all ordered into a very methodical twisted, ladder-like structure known as a double helix. In essence, genes are made up of molecules. And what are molecules? Molecules are arrays of two or more atoms. For instance, a sugar molecule, like the one in DNA, is made up of a combination of carbon, oxygen, and hydrogen atoms.

Carbon, hydrogen, and oxygen atoms, nitrogenous bases, phosphates: these are the ingredients in the recipe for making a human being. Stored in the particular arrangement of these atoms exists all the information necessary to create a person's entire physical make-up, all accounted for before that person is even an embryo, not to mention born. A person's sex, skin and eye color, height, vision, hearing, proclivity for high blood pressure, asthma, diabetes, schizophrenia, Alzheimers, and allergies, as a few examples, all exist within this first fertilized cell - one's physiological life story told from the moment we are conceived.

The sperm and the egg meet to create one very "informed" fertilized cell. Stored within this first cell are instructions to divide. Once this occurs, the emerging person exists as two cells, each which contains all the information necessary to create a fully developed human being. These two cells will now reproduce, and so on, and so on, until a cluster of cells form. Stored within each of these cell's chromosomes is information that will now instruct the cells to start producing even more specialized ones, such as nerve cells, blood cells, and muscle cells. With the emergence of these specialized cells, the unborn embryo will continue to differentiate and grow within the mother's womb, until nine months later when it is ready to be born.

So all our traits are, generally speaking, predetermined from the moment of our conception. But what exactly are traits? Traits are those characteristics that distinguish not only one species from the next, but each individual from all others within its own species. And from where do these traits emerge? They originate from information stored within an organism's genes, that unique arrangement of atoms that make up an organism's chromosomes.

For example, the fact that all fish have gills would imply that somewhere in a fish's chromosomes lies a gene or group of genes that instructs the developing embryo to produce gills. This is not just true of the fish's gills but of every single physical characteristic the fish possesses. As no trait can develop spontaneously, as of its own volition, this means that for every trait fish possess there must exist some corresponding gene responsible for its emergence. Unless we are to believe that all fish have gills as the result of some incredible coincidence, we must accept the evolutionary, that is, the genetic explanation for such a phenomenon. If fish possess gills, there must exist "gill" genes. If a fish is equipped with fins, there must exist "fin" genes, and so on and so on, until every single physiological characteristic is accounted for. In this way, the developed animal is a composite of traits that correspond to information stored within the animal's genes - once again, information already established from the moment of that animal's conception.

As each species possesses its own unique set of traits, each species must possess its own unique set of genes. The fact that fish possess gills means that the molecular arrangement of their genes must be different from a creature that has no gills. The fact that *all* fish possess gills means that gill genes must exist in *all* fish DNA.

Since each individual that emerges from a sexually reproducing organism is formed from a unique blend of its two parent's chromosomes, each individual represents a slight variation from one to the next. In this way, though all fish may possess gill genes, each fish's gills will, in some slight way, vary from one individual fish to the next.

The same is true for humans. Though we all possess genes that instruct our bodies to develop two eyes on our heads, each person's eyes are slightly different. This is true for every characteristic we possess as a species. Whether we are discussing one's height, sense of hearing, skeletal or facial structure, the constitution of one's heart, kidneys, or immune system, each part of us varies in some way from one individual to the next. In a sense, every single part of us, from every cell to every organ, is as unique to each individual as are one's fingerprints, which, though we all possess them, no two are exactly alike.

Regarding these slight variations each individual possesses - in the constant competition of life, those creatures whose variations are best suited or adapted to their surroundings are at a considerable advantage and are, therefore, that much more likely to survive. Those forms more likely to survive will, in turn, have a greater chance of reproducing. Those that have a greater chance of reproducing will, consequently, have a greater chance of passing their genes - along with their advantageous traits - on to future generations.

Just as each individual within a species varies in some way, so does the entire gene pool of each generation of that species. Just as no two individuals are ever alike, neither is the gene pool of any two generations of a given species. Because each generation is put through another screening of natural selection, each generation will most probably be that much better suited to its environment. In this way, life is in a state of constant flux, each species constantly maturing and evolving with each passing generation.

Let me provide a more concrete, though hypothetical, illustration of how this process of natural selection works: Imagine a place where the land is flat, lush with plant life, covered in grass and trees. Roaming this land is a hypothetical three foot tall, horse-like creature I will call the nequus. A male nequus and a female nequus mate and have three baby nequuses. Given the way the two parents' genes recombine, the three children will all vary from one another. Regarding, for example, the offsprings' heights, the average height of the three combined should *most probably* equal the average height of the two parents. In this way, one of the three offspring *should* turn out to be a bit smaller than its parents' average height, another *should* turn out to be approximately the same as its parents' combined average height, while the third *should* turn out to be a bit taller than both of its parents.

Back to the nequus plains: Imagine a geological event were to take place that would transform this once lush region into an arid one. Amid these new conditions, imagine much of the plant life has died. The nequuses, which are herbivores, suddenly find themselves in fierce competition for what remains of their now dwindling food supply. Unfortunately, the average nequus, which is only about three feet tall, can only reach the bottom branches of its region's trees, much of which have already been eaten.

Back to our offspring: Of the three, which is most likely to live long enough to reproduce and, consequently to pass its genes onto future generations? Obviously, the tallest. Why? Because it can reach the leaves of those higher branches that the majority of its starving species cannot.

Let's now imagine that the taller nequus lives long enough to mate with another that is, let's say, average in height. Between the two, they produce three children. As the male is taller than its father was, the average height of this new litter of offspring will *most probably* be greater than those of its own. Over a period of several generations, this process might continue until the average nequus is now, let's say, four feet tall compared to its three foot tall ancestor. In this way, each species is constantly being modified to most effectively meet the demands of its ever-fluctuating physical environment.

In the case of the imaginary nequus, should the drought and consequent food shortage continue, the forces of natural selection will continue to weed out those least suited to survive these conditions and to preserve those that are best.

Perhaps after a period of ten million years of such natural selection - what would amount to the passing of approximately a hundred thousand generations - the average height of a nequus may have grown to be ten feet tall, making it resemble something more like a giraffe than its predecessor, which was more like a horse. In essence, what used to be a nequus has now evolved into a different species with a new sequence of genes. Apparently, necessity is the mother of selection.

To provide an actual example of how environmental pressure can alter a species' physiology, I'll now refer to the real-life case of the Biston Betularia or what is more commonly known as the Peppered Moth. During the 1800's, it was noticed that this once predominantly white mottled moth had, within a very short period of time, evolved into a much darker variety. Originally, the lighter variety had spent much of its time resting on similarly colored trees. In this way, the peppered moths were pigmented in such a way that they matched the coloration of the trees they rested on, thus making it much more difficult for predatory animals to see them. With the advent of the industrial revolution, however, residue from nearby factories covered the forests with dirt and soot, consequently, darkening the surface of the trees. Because the white moths, which represented the majority of the species, could now be more easily sighted by predators, they became more likely to be eaten. By contrast, the darker variety of the moth's population, those which previously represented a minority, were now much less likely to be seen by predators and, consequently, that much less likely to be eaten. Because they were less likely to be eaten, the darker variety were now more likely to survive and therefore to pass their genes on to future generations. As a result of this sudden change in the environment, the population of the moths had quickly shifted so that the species' darker strain, once the minority, now came to represent its majority. And so, within just a few generations, the entire peppered moth population had been modified due to an environmental pressure.

Another aspect to the theory of evolution involves a process known as genetic drift. To illustrate this process, imagine that due to overpopulation certain members of a species find themselves having to migrate to a new area in search of new sources of food. For instance, ten finches among a community of tens of thousands migrate to a nearby island in search of food. Since these ten finches can never represent the exact genetic mean of their species, should they reproduce, they will, in turn, be creating an entirely different strain based on their own particular physical characteristics. In a sense, these ten pioneer finches would represent the founders of a whole new, slightly different, genetic strain. Because of the pioneer group's slight genetic variance from the mean of its original population, this new strain might, in time, eventually come to represent a whole new species. As a matter of fact, this is exactly what Charles Darwin discovered when he went to the Galapagos islands to study the various species of finches as they existed on each of its separate islands. Through his observations, Darwin noticed that the finches from each of the Galapagos' separate islands seemed to represent a unique strain. It was from these observations that Darwin first conceived of his theory of evolution.

Returning to my study of Homo sapiens: With the advent of humans, there came a whole new panoply of specifically human sciences. The first of these human sciences was anthropology.

This dealt with matters concerning the social, behavioral, and anatomical evolution of those advanced primates, the hominids, all the way up until about ten thousand years ago when humans reached what is referred to as the neolithic stage of their existence. What separates neolithic humans from their biologically identical ancestors was the discovery of agriculture. Before the neolithic period - during what is known as Man's paleolithic period - these more primitive humans drifted across the globe in nomadic tribes, constantly moving about in search of new food supplies.

But humans possessed an evolved brain and, over time, began to notice patterns in their physical environment. Unlike any other animal that came before it, humans could recognize, for example, that where a plant's seed fell, a new plant would often emerge. When the first humans made this connection, about twelve thousand years ago, it enabled them to imitate nature by planting their own crops. With the advent of agriculture, the human animal began settling into stationary communities (usually near some river, which allowed for a constant water supply as well as a means of transportation). Furthermore, by noting the manner in which the other animals reproduced, humans learned to herd these animals into ranches where they could control their meat supply to supplement their supply of fruits and vegetables. The combination of these two events is referred to as the agricultural revolution. It is referred to as a revolution because of the immense impact this discovery had on our species. For the first time in our species' history, humans could regulate their own food supply. No longer needing to devote all of their time to searching for their next meal, humans could afford themselves some *extra* or what we call leisure time. With all this additional time on their hands, humans had the opportunity to direct their energies to self-expression [the arts], play [sports], as well as the pursuit of wisdom and knowledge [philosophy and science].

As some of these agricultural settlements began to flourish, other peoples began to migrate to them hoping to reap the benefits of these new establishments. In time, some of these settlements began to expand in size and population. It was here, in these first cities, where humans from a variety of cultures first congregated in order to exchange goods as well as ideas. This marked the dawn of a period in our species' history known as the urban revolution. As these cities continued to grow, humankind's first civilizations arose.

As time went on, civilizations rose and fell. Without reciting the histories of all the various civilizations, suffice to say that this process continued until I, Matthew Alper, find myself here at the tail end of the twentieth century in a modern-day megalopolis known as New York City.

Now I make no claim that science could explain everything. Sure, there were parts of the physical universe that were better understood than others. Sure, there were whole fields that were, in many ways, still incipient and, consequently, theoretical in nature. Sure, there were still mistakes to be made, details to be reworked and revised. Generally speaking, however, the scientific interpretation of the universe always remained true to its method, one that has given us nuclear energy, electron microscopes, electric light, and antibiotics, as examples of its awesome capacity. Here were technologies I knew, as a fact, worked; things that took a great deal of scientific research to create; the same type of research, the exact same methodology that was used to account for the aforementioned history of the entire physical universe. THE PROOF WAS IN THE PRODUCTS. If I could rely on the scientific method to create such wonders as space shuttles, gene therapy, nuclear power, and microwave ovens, then why shouldn't that same methodology to be able to explain the origin and evolution of the entire physical universe as well as of terrestrial life. How else could science have so successfully mastered and manipulated our physical world if it didn't understand its very nature?

Science had accounted for the approximately fifteen billion year history of the entire physical universe, from its origins to its present state and all without the aid or assistance of any spiritual power - COSMOLOGY WITHOUT GOD! Science had been equally able to account for the approximately three and a half billion years of organic evolution, also without the aid or assistance of any transcendental force or entity - THE ORIGIN AND EVOLUTION OF LIFE WITHOUT GOD! No longer was either life or the universe contingent upon the existence of some god. Not to say this meant that God didn't exist, but let's just say, it reinforced the possibility.

No longer would I have to ask such questions as, "If there is no God, then how is one to explain the origin of life?" Or "Without

God, how did the Earth, the Moon, the Sun, and the stars all come to be?" No longer would I have to look down at my own physical body and not understand the origin, evolution, nature, and mechanics of my own being.

All this, science had done for me. First it rescued me from the clutches of mental illness, and now it had made the universe comprehensible to me. And yet, there it was, taunting me as much as ever, that incessant longing, that gnawing pang to know not *how* I or the rest of the universe worked but *why*? There it was, still looming over me, as oppressive as ever, that ever-weighty problem of the purpose of my existence. Why was I here? What was my point? As always, underlying this question was the equally elusive problem of God. Only knowledge of God could resolve the ultimate question of my existence. And yet, how was it that amid all of this glorious information the sciences had yielded, it couldn't offer me any explanation, whatsoever, regarding the nature of God's existence? Was God simply incomprehensible to us? Or was there a scientific explanation, only no one had discovered it yet? What pattern in nature, what empirical observation, I wondered, might possibly help to reveal the nature of God's existence to humankind? Then again, even if there was a solution, might it not lie beyond our reach - a problem meant to oorment and tantalize us until the end of time?

Regardless of whether the problem was answerable or not, all I knew was that, spiritually speaking, I had yet to be satisfied. The search would just have to go on.

FOUR

KANT

"All that I experience is psychic. Even physical pain is a psychic event that belongs to my experience. My sense – impressions – for all that they force upon me a world of impenetrable objects occupying space – are psychic images and these alone are the immediate objects of my consciousness. My own psyche even transforms and falsifies reality, and it does this to such a degree that I must resort to artificial means to determine what things are like apart from myself. Then I discover that a tone is a vibration of the air of such and such a frequency, or that a color is a wave-length of light of such and such a length. We are in all truth so enclosed by psychic images that we cannot penetrate to the essence of things external to ourselves. All our knowledge is conditioned by the psyche which, because it alone is immediate, is superlatively real. Here there is a reality to which the psychologist can appeal - namely, psychic reality." — Carl Jung

So far, my search for knowledge of God had been directed outward, onto those objects that constituted the entire physical universe. I had studied the physical nature of atoms and molecules, of planets and stars, of organic and inorganic compositions of matter. And still, no matter where the astronomers had pointed their telescopes, or which specimens the biologists had placed under their microscopes, or even which particles the atomic physicists had split asunder, not one had ascertained anything resembling verifiable knowledge of any spiritual reality or God. And so, in order to complement my investigation into the natural/physical sciences, I was simultaneously studying that discipline known as philosophy.

Though its roots mean the love of knowledge, philosophy, as I saw it, represented the study of the ultimate nature of reality. What, if anything, can be said to be real? What, if anything, can be said to represent truth? In essence, what is reality?

The ancient Greeks, who are generally recognized as the

founders of western philosophical thought, believed that in order to understand the nature of reality one had to first understand the nature of those objects that filled the vast physical universe. The ancient Greeks, therefore, conducted their investigation into the nature of reality by studying those objects which existed around them, essentially those objects they could perceive with their physical senses. Whether it meant studying something as seemingly simple as a table or a chair, or perhaps something seemingly more complex such as a living organism, or perhaps something of which it was just harder to obtain knowledge simply because it existed beyond our physical reach, such as the planets and the stars, the ancient Greeks sought to understand the underlying nature of reality through those objects that existed around them. Of what, for instance, are such objects made? In what ways are the various objects in our world similar? In what ways are they different? These were the types of questions the ancient Greeks felt needed to be resolved if the true nature of reality was ever to be fathomed.

Similar to the Greek method, this was how I, too, had been conducting my own personal investigation - by studying the nature of those material objects that permeated the fifteen billion year history of the entire physical universe. This was the method by which I, too, sought to comprehend the nature of ultimate reality, a problem I presumed would, once resolved, lead me to an even more comprehensive knowledge of spirit and God. I, like the Greeks, had been looking outward for answers, that is, until I came across the work of the eighteenth century German philosopher, Immanuel Kant.

Since the ancient Greeks first introduced this particular method of inquiry (of looking into the nature of things external to them), this represented the predominant trend in all human science and philosophy up until the eighteenth century, when Immanuel Kant arrived on the scene. In his book, *Critique of Pure Reason*, Kant had made one of the most revolutionary leaps in the history of all human thought by suggesting that in order to understand the true nature of reality, we need to redirect the focus of our inquiry from outwards to within. Kant proposed we do this by studying, not the nature of those physical objects around us, but rather the manner in which we perceive those objects. Rather than looking outward for our answers regarding the ultimate nature of reality, we first need to look inward, into the nature of that which is doing the perceiving.

Take, for example, an apple. By what means, Kant asked, do we come to have knowledge of an apple? The answer: through information we acquire through our physical sense organs. Through the absorption of reflected photons of light as they fall on our retinas, we *see* the apple. Through molecules the apple releases into the air, we *smell* it. As its chemistry dissolves on our tongues, we *taste* it. Through the pressure of the apple's contours against our skin, we *feel* it. Only after we process this medley of electrical signals, of sensations, do we come to possess knowledge of the apple as a whole. By this reasoning, however, it isn't the apple "itself" we come to know but rather the apple "as-we-perceive-it." According to Kant, it is impossible for humans to ever possess objective knowledge of any "thing-in-itself." Such absolute knowledge of things, Kant referred to as *noumena*. Instead, Kant suggested, we can only know "things-as-we-perceive-them" or that which he referred to as *phenomena*. Consequently, everything we call knowledge is subjective, relative to the manner in which we process and, therefore, interpret reality.

Kant's ideas actually represented the evolution of the thoughts of the seventeenth century English philosopher, John Locke. According to Locke, human beings are born as clean slates - a "tabula rasa" as he phrased it, our minds "empty tablets capable of receiving all sorts of imprints but have none stamped on them by nature."

Almost a hundred years later, Kant wondered, how was it possible that this jumble of data we receive through our sense organs could spontaneously arrange themselves in such a way as to yield coherent information? How is it that this vast stream of stimuli we are constantly being bombarded with all manage to fall into place in such an intelligible manner as to provide us with practical knowledge of our world? According to Locke, this process occurs automatically. Not so, contended Kant.

According to Kant, there was just no way this multitude of sense-impressions could arrange themselves in such an effective manner of their own volition. Apparently, the human mind, Kant asserted, must be born, not as a clean slate, but with inherent mechanisms that work to organize the vast medley of information our sense organs constantly impart to us. Without such built-in processing mechanisms, we would experience reality as an unintelligible jumble of sense-experiences. It is therefore necessary, Kant argued, that there exist some inherent structure of the mind that functions to order the profusion of sensations we receive. The human mind is therefore not some passive organ, as Locke would have had us believe, waiting for experience alone to shape and define us, but rather an active one whose function is to order the multitude of information our sense organs impart to us.

Two of the ways Kant suggested that humans inherently process information was temporally and spatially. According to Kant, humans are equipped with built-in processing mechanisms that work to provide spatial and temporal order to our experiences. As we are creatures that live through time and in space, it would be impossible for us to make sense of our experiences without such internalized cognitive processors. Kant therefore maintained that we are born with physiological mechanisms that act to order our perceptions into spatial and temporal perspective. For every object we perceive, we have an inherent tendency to define that object as existing either here or there, now or then. Accordingly, space and time are therefore not things we perceive "in-themselves" but rather they represent two inherent modes of perception, that is, two of the many ways our species instinctively processes the vast array of information we receive. Our comprehensions of time and space are, therefore, not concepts we learn through experience but rather represent two of the means through which we inherently perceive and, consequently, interpret reality. Because our comprehensions of temporal and spatial dimensions exist in us as an inherent part of our cognitive processing, Kant suggested we possess such awarenesses "a priori," meaning prior to our actual experiences of them.

As I contemplated Kant's ideas, I recalled the work of the developmental psychologist, Jean Piaget. Based on a series of experiments he had conducted, Piaget concluded that a child cannot distinguish certain proportions of time and space before reaching a certain age, which he referred to as the stage of concrete operations. Piaget found that before this stage in our natural cognitive developments, not only are children unable to distinguish proportions of time and space, but they cannot even be taught to comprehend such concepts. To demonstrate this, Piaget placed two glass beakers before a number of children. Though one of the beakers was short and wide and the other tall and narrow, both were equal in volume. When asked which of the two beakers would hold more liquid, it was the children's inclination to believe that the tall, narrow one would. In order to show that the two beakers were equal in volume, Piaget filled the short, wide one with water until it was filled. He then poured the contents of this first beaker into the tall, narrow one. Because the two beakers were equal in volume, as the short, wide one emptied, the tall, narrow one became filled. What this should have clearly demonstrated was that both containers were equal in volume.

After the demonstration was complete, Piaget again asked the children which container held more liquid. On this second questioning, children ages seven and up almost invariably answered that the two beakers held equal amounts, while those who were younger still believed that the tall, narrow one could hold more. What this demonstrated was that not until children reach a certain age can they even be taught to comprehend certain spatial proportions.

Based on this data (along with the results of similar experiments conducted on the comprehension of temporal dimensions) Piaget theorized, just as Kant had years before, that there exist certain concepts we comprehend as a part of our inherent cognitive processing as opposed to things we simply learn. Apparently, the human animal goes through a series of developmental stages through which we acquire our various cognitive capacities. The fact that Piaget had proven that our ability to distinguish temporal and spatial relations emerges in all humans at approximately the same age suggested that such aptitudes represent an inherent part of our species' natural cognitive development, something Kant had proposed approximately two centuries earlier.

So perhaps Kant was right. Perhaps we do inherit our capacities to comprehend dimensions of time and space. What this implied was that humans are born with certain inherent modes of perception, ones that frame and define our perspectives of reality. Was it possible, I now wondered, to apply Kant's principles to the subject of human spirituality, that is, to my own personal quest for knowledge of God? Had I been wasting my energies trying to ascertain knowledge of God's existence through the study of those objects external to me when, instead, I should have been studying the nature of my own internal cognitive mechanism? Perhaps I needed, as Kant had done, to invert the nature of my quest from outward to within. Perhaps the solution to the problem of God's existence lay not "out there" but within the workings of my own mind or, as my biopsychology texts would have it, my brain. Perhaps there was something to this.

...Just perhaps!

FIVE

<u>GOD AS WORD</u>

"In the beginning was the word, and the word was with
God, and the word was God."   — John, 1:1

I was now thirty-one years old, approximately ten years since I
had begun my formal exploration into the natural sciences in the hope
that it might yield some small knowledge of God. In this time, I had
learned about the fifteen billion year history and evolution of the
physical universe. I learned how the universe was born and of its consequent expansion; how the forces of gravity would one day overwhelm the universe's expansionary thrust causing all of its matter to
once again collapse upon itself, thereby reuniting all matter and energy into one condensed single point, the same as it was the moment
before the last Big Bang occurred; how, at this time, yet another
explosion would occur that would cause the whole process to start all
over again; how this process of expansion, equilibrium, and contraction; expansion, equilibrium, and contraction would repeat itself over
and over, ad infinitum, like a great pulse in space that would beat
until the end of time.

Science had taught me the origins of matter, the atoms, the stars,
the planets, the earth, of organic matter, of life, of humankind...of Me.
For almost every single physical phenomenon I sought to understand,
I found that science had already gone there. And yet, with all its wisdom, its passion, its inquiries, and investigations, science could not
provide me with the vaguest knowledge of God. What was this entity that could elude such men of will and genius, such men who
brought us laser beams, space ships, and nuclear energy?

Would anyone ever resolve this God-forsaken riddle? Where was
God? Where was He hiding? How was it that we all knew who He
was, that we could all talk about Him, that He played such an important role in all of our lives, and yet not one of us had a clue as to His
whereabouts? What mischievous imp was this that He should create
us to believe in Him and then stand tantalizingly out of our reach?
Why not just make Himself known to us already? What, after all, was

the big secret?

So here I was, years later, as uncertain as ever as to the point or purpose of my existence. The only difference between myself now and before was that at least now I was armed with an arsenal of scientific information, none of which, to my dismay, had imparted the vaguest knowledge of God. Was it that I had yet to place all my new-found data into its proper perspective? Or was it simply that God existed beyond the scope of the physical sciences, beyond the scope of human reason? Whatever the case, I finally decided to apply the scientific method to my own search for knowledge of God. I would take a step back and review the question in an organized and methodical way...the scientific way.

And so, I drew up a review sheet, one that would address the question: With everything I had thus far learned in life, what, if anything, could I possibly say that I knew of God's existence? Had I ever seen God or witnessed anything that could prove he existed? No. Had I ever heard of anyone else witnessing the divine? Sure. People made claims all the time to have witnessed a statue bleeding or some other such miracle. Only, when was the last time that such an account had been validated, authenticated, or substantiated by scientific method? The answer, of course, was never. Not once in my lifetime had anyone captured one single miracle on any reliable medium (Just think of the coverage the parting of the Red Sea would have gotten if it took place today). Why was the past so replete with divine interventions and miracles while the present contained none? When was the last time the church had sanctioned or endorsed a miracle? Certainly none I had heard of, not at least in the last few centuries. Why was this? Why was it that all the renowned and celebrated miracles had all occurred in ancient times? Why not today? Had God simply abandoned us since then? Or did it have something to do perhaps with the fact that it was approximately that many years ago that the scientific revolution took place, something to do with the fact that once the scientific method took root all such claims of miracles were now placed under science's strict and unswerving scrutiny?

Now, with the advent of scientific culture, if a person were to claim to have witnessed a miracle, he would have to be able to prove it. No longer could any drunk walk into town claiming to have beheld God or some other such miraculous event without having to answer to a body of scientific investigators, all asking a battery of questions, cross-referencing answers, seeking physical evidence, conducting experiments, hooking him up to electro-encephalograms, etc. Now, with the advent of scientific culture, if the lone drunk walked into town claiming to have witnessed the divine, he faced the potential risk of being ridiculed, if not sedated and prescribed anti-psychotics.

Back to my own personal dilemma, the situation was simple. Unless I was presented with some form of tangible evidence that could verify God's existence, there was no way I was going to believe that He/She/It existed.

But why? Why this need to conduct such a quest in the first place? Why continue to submit myself to this frustrating investigation, following one false lead after the next? At this rate why not just spend the rest of my life searching for a unicorn or some other such fantastical creature? Why restrict my aimless quest to this one particular imaginary being? Why this exclusive obsession with this entity we call God? It was as if the need to comprehend an absolute being was somehow instilled in me. Just as I was driven to seek food, shelter, security, and love in my life, I was driven to possess spiritual certainty, driven to search for knowledge of God. But why? There must have been some reason for this. Nothing springs from nothing. As science had taught me, everything that occurs in the physical universe has its physical cause. There had to be some reason, some physical cause that this particular obsession persisted in me the way it did.

Perhaps I was insane. What else could explain such an abstract compulsion? Perhaps this was the answer. Only if I were insane, then so was almost everybody else on this planet, for more peculiar than the fact that I was so preoccupied with this particular obsession was the fact that so was almost everyone from every world culture from the dawn of my species. This was not my own personal idiosyncrasy, but one that, I, oddly enough, shared with nearly every person from every culture I had ever experienced, heard of, or read about, dating as far back as to the origin of my species. What kind of bizarre coincidence was this? Sure, everybody had his or her own eccentricities, but why was it that we all shared this particular one?

Some people explain human behavior as the sum of one's life experiences. As we all live such unique lives, how might this possibly explain the fact that every culture from the beginning of our species has believed in the same thing? How is it that every culture has maintained a belief in a spiritual/transcendental power or being, a god? How was it that people from every walk of life, every race, age, sex, and class, all shared a belief in some form of a spiritual reality, a god, a soul, and an afterlife? How odd that if I were to sit down with another person of any race, age, sex, class, or culture, I would be able to hold a conversation (providing, of course, that we spoke the same language) concerning the nature of some spiritual/transcendental reality, a God or gods, a soul, and the possibilities of an afterlife. Perhaps this was proof, in itself, that God existed. What else could explain the fact that billions of people from every generation, from every culture - even isolated ones - had all pondered these very same notions? Unless this was the result of some vast and incredible coincidence, some force, drive, or internal pressure must have been responsible for this strictly human phenomenon.

And so, I stepped back once again and asked myself, "What can I say with near certainty that I know of God?" when suddenly, in one radiant and Archimedic moment, it dawned on me. As plain as the nose on my face, right there on my computer screen before me lay the one small but certain fact I had been searching for. As insignificant as it may have seemed, I realized there was something I could finally say I knew of God's existence with near absolute certainty: Simply, **GOD WAS A WORD!** There it was, spelled out on my computer screen before me, my first empirically verifiable fact regarding God's existence. I could read it, write it, hear myself say it. In Braille, I could even touch it. No doubt about it - God, I could say, with empirical certainty, was a word.

Should I doubt my own sanity, I could always look up the word in any dictionary. If this wasn't enough, I could always go anywhere in the world and ask those around me if they were familiar with the concept of a supreme spiritual/transcendental being, a god. Who could deny that at some point in his or her life he or she hasn't considered the existence of some spiritual element in the universe? What functional adult has not, at some point, contemplated the concept of some transcendental force or being? Not even an atheist could make such a claim.

So God was a word, a word that represented the concept of a transcendental/spiritual force, power, or being. More compelling yet, here was a concept for which every culture from the beginning of my species, no matter how isolated, possessed their own symbol or word.

And what exactly did science have to say about words? Where, for instance, did they originate? One place: the human mind. Only "mind" seemed an ambiguous term. In nearly all of the religious/ philosophical texts I had ever read allusions were constantly made to a mind/body dichotomy, implying the two were separate entities, two distinct agents. Mind intimated that consciousness possessed some transcendental quality. It allowed for the existence of a spiritual component in us. As science had never confirmed the existence of such a component, from hereon, I would only use the word "brain." Minds, science could not verify. Brains, they could. No differently than we would view a heart or a kidney, brains were viewed as being one hundred percent spirit free, purely physical/organic/mechanical in nature.

So God was a word that, like all words, originated from within the workings of the human brain. Before humans existed, there were no such thing as words. Words originated, as did the concept of God, with our species. Now, if brains were strictly biological in nature, and the word "God" originated from within that same organ, then perhaps the concept of God was somehow inextricably linked to our biological natures. Could it be that the concept of God was somehow a product of my species' inherent cognitive processing, the manifestation of some innate "spiritual" mode of perception? Was it possible that the solution to the problem of God's existence lay not "out there" but rather buried somewhere within the recesses of the human brain?

The one thing I could now say of God with any empirical certainty was that God was a word, which, like all words, was generated from within the human brain. This meant the only fact I now possessed regarding the nature of God's existence came, not from something I had perceived from without, but rather from something that had been generated from within, more specifically, from within the workings of my physical organ, the brain - and not just my brain but from those of almost every single person from every single culture dating back to the dawn of my species.

Trying to decide where to best take this notion, I remembered the position held by the sciences that if a behavior was universal, that is, if it was realized by every member (or in the case of humans, among every culture) of a given species, then it must represent an inherent characteristic of that organism's physical nature, that is, a genetically inherited trait. And so, as surely as all human cultures have spoken a language or engaged in sexual reproduction, all cultures have practiced a belief in some form of a spiritual reality. Did this then mean that our perceptions of a spiritual reality, of a god, might also represent the effects of a genetically inherited trait? And if so, how might I possibly prove such a thing?

SIX

UNIVERSAL BEHAVIORAL PATTERNS

"It is universally acknowledged that there is a great uniformity among the actions of men, in all nations and ages, and that human nature remains still the same in its principles and operations. The same motives always produce the same actions. The same events always follow from the same causes."   - Hume

"I will analyze the actions and appetites of men as if it were a question of lines, planes, and solids."   - Spinoza

"One needs to look near at hand to study men, but to study man one must look from afar."   - Rousseau

Whenever there is a physical characteristic that is shared by every individual among a given species such a characteristic must represent a genetically inherited trait.

The fact, for instance, that all Monarch butterflies possess the same colorful pattern on their wings implies that such a universal physical feature as this must represent an inherent characteristic of that species, a genetically inherited trait. One could challenge this assumption, but how else might they explain the uniformity of the Monarch's design? Are we to imagine that all Monarch butterflies possess this same complex pattern on their wings as the result of some vast coincidence? As if Monarch butterflies could have been born with any pattern or combination of colors on their wings, only it just so happens that, by pure chance, they all turned out the same? Hardly! No less of a coincidence than the fact that all fish have gills, all Monarchs possess the same elaborate design and uniform coloration on their wings. Apparently, the Monarch's unique display exists as the result of information encoded in the genes of each member of its species.

The same can be said of all universal characteristics possessed by a given species. Whether we are discussing a butterfly's wings, a rat's tail, a cat's whiskers, a fish's gills, or a human's brain, each represents a universal physical characteristic that emerges as the result of information stored within that species' genes. We could therefore say that, as a rule, for every physical characteristic common to every member of a species, there must exist genes that promote the emergence of that trait.

Not only does this rule apply to universal physical features but to universal physical functions as well. For instance, all humans grow hair. The fact that growing hair represents a universal characteristic of our species would suggest that this, too, must represent a genetically inherited trait. Because growing hair represents a function, this would imply that our species must possess some specific gene or set of genes that instructs our developing bodies to forge a physiological site in us, some organ or mechanism from which hair growth will be generated. In this particular case, this site is represented by our hair follicles. Unless we are to believe that hair "magically" appears from our skin, it is necessary that some physiological site exist from which hair is produced.

What this suggests is that for every universal function a species possesses, whether it be an organism's capacity to smell, taste, hear, see, breathe, pump blood, reproduce, move, secrete hormones, etc, three things must be true: one, all functions that are universal to a given species must represent an inherited characteristic of that species; two, for each of these inherited characteristics, there must exist some underlying gene or group of genes responsible for the emergence of that trait; and three, there must exist some very specific physiological site from which that function is generated.

As another example of a universal function, let's examine the movements of the planarian, a microscopic creature belonging to the family of flatworms. Planarians do not have brains but instead have several longitudinal nerve cords that run the length of their tiny bodies to a head where these few nerves converge. Rather than calling this convergent cluster of nerves a brain, it is referred to it as a ganglion, constituting a relatively simple central nervous system.

Every member of the planarian species has a distinct tendency to position its body so that it faces the light, a phenomenon referred to as phototactic behavior. That all planarians engage in this particular behavior implies that it represents a universal characteristic of the species.

There are three possible reasons that all planarians respond to light in this particular way, the first being they were taught to do so by others of their species. In other words, perhaps all planarians position themselves in the direction of a light source as the result of a learned behavior. The problem with this explanation is that, should we isolate any one individual planarian from the moment of its conception, allow it to develop in isolation to its adult stage, and then place it in an environment with a light source at one end, it will invariably turn in that direction. This would imply that phototactic behavior is not one that needs to be learned by this species.

The second possible reason all planarians orient themselves in the direction of light is that they *want* to, that is, they do it as an act of free will. As we can never truly know what a planarian is "thinking," we can never know whether this is the case or not. Nevertheless, if planarians did have the wherewithal to make such free and voluntary decisions, what are the chances that every single one of them would always choose to act in the same exact way? What would be the likelihood that every single "free thinking" planarian would independently *choose* to orient itself towards the light all of the time? Wouldn't it be reasonable to presume that there would be just one planarian that would "choose" to turn away from the light, even if just some of the time? Are we to believe that all planarians always engage in phototactic behavior as the result of some vast coincidence, as if they actually do possess free will, and that any day now they might all suddenly change their *minds* and decide to shun the light? Again, highly unlikely. That all planarians always turn towards the light leads me to believe that this is not a case of either free will or coincidence.

The third possible reason that all planarians demonstrate this propensity towards phototactic behavior is that infused within the planarians' ganglia exist genetically inherited instructions that compel each individual within the species to respond to light in this particular way.

So which one of these three possibilities represents the true reason planarians orient themselves towards light? Well, as I have already suggested, it's not very likely this behavior is a learned one. It's even less likely that it's the result of coinciding acts of free will. This leaves us with the third possibility which states that phototactic behavior represents a genetically inherited response to a specific stimulus, what is otherwise known as a reflex.

Nonetheless, science cannot base theories on the process of elimination. If we wish to speculate that planarians orient themselves towards the light because they are physiologically "wired" to do so, we had better be able to prove this.

Planarians perform this feat of orienting themselves towards the light by shifting themselves until their two light receptors (what we would call their eyes) are equally stimulated. When the two receptors are equally stimulated, the planarian stops moving, thus guaranteeing that it will directly face the light. In experiments performed on planarians, it was found that "If two equally bright lights a short distance apart are placed near the planarian, the animal will orient itself toward a point midway, thus attaining equal stimulation of the two eyes."[4]

That we can manipulate the planarian's behavior in such a way implies that this organism responds to light neither as an act of free will nor as a learned behavior, but rather as the result of a physiologically generated involuntary response to a specific stimulus. In other words, planarians orient themselves in the direction of light because somewhere in their ganglia there exist specific nerves which compel them to do so. Similar to the way we can wire a mechanical toy or robot to turn towards the light, nature has "wired" planarians with this same propensity. We could therefore presume that somewhere in the planarians' ganglia they possess a phototactic function, a nerve impulse that prompts them to respond to light the way they do. Should we locate and then cut this particular nerve in the planarian's ganglion, it would most likely lose its capacity to perform this phototactic reflex.

That all planarians exhibit this same impulse to respond to light in such a specific way implies that such a universal behavior must constitute a genetically inherited trait. In other words, **behaviors can represent genetically inherited traits**. No differently than Monarch butterflies inherit their unique wing design, planarians inherit this phototactic reflex.

So what of universal behaviors that exist among the more advanced species? Is it possible we could apply this same principle to every other form of life as well? Consider, as another example, the fact that all honeybee colonies construct their hives in the same hexagonal design, regardless of whether or not they've ever had exposure to another honeybee colony. If bee larvae were to be removed from their colony and raised under artificial conditions, they would still emerge as adults to construct their hives in this same hexagonal fashion. If we were to apply the same principle that we did to the planarians to honeybees, it would imply that the bees construct their hives in a hexagonal shape as the result of a neurological impulse, an inherited reflex, something genetically preprogrammed into their brains. We could therefore say that bees construct their hives similarly to the way that planarians respond to light, as the result of a genetically inherited reflex.

Moving right along the evolutionary ladder, how about the fact that all adult peacocks display their feathers when exposed to an aroused peahen? Wouldn't the fact that all peacocks respond to the same stimulus in the same exact way imply that such behavior must represent a genetically inherited reflex? (The only difference being that when we speak of animals belonging to the subphylum of vertebrates, in this case a bird, we tend to refer to such reflexes as instincts.) This would further imply that peacocks must possess some "display" mechanism somewhere in their brains, one meant to promote the act of courtship and, consequently, reproduction. Should we locate and then remove this particular part of the peacock's brain, it's more than likely it would no longer be able to appropriately respond to the stimulus of an aroused female by flaunting its feathers.

Continuing up the evolutionary ladder, all Eastern Mountain gorillas engage in species-specific play behaviors, courting and reproductive rites, foraging techniques, displays of territoriality, and aggression, as a few examples. How is it possible that all troops belonging to this species, regardless of whether they've been exposed to one another, engage in such similar behaviors? Are we to believe that the species is psychic and can telepathically communicate their behaviors to other troops across the plains? Or is it because since all Eastern Mountain gorillas exist as a part of the same species and therefore possess such similar genes, they are programmed to behave in such similar ways? Though a peacock and a female gorilla would never be able to effectively court one another, a male and female gorilla from any two troops could. That Eastern Mountain gorillas demonstrate such universal behaviors again implies that such behaviors must represent genetically inherited instincts. Just as all planarians turn towards the light, all gorillas engage in certain species-specific play, grooming, foraging, and courtship behaviors. Does this then mean that gorilla behavior, similar to planarian, bee, or peacock, exists as the result of a series of preprogrammed reflexes?

> "Should we then regard reflexes as the fundamental behavioral units? In a sense, yes....And it is true that there is no difference between simple reflexes and more complex reactions; every conceivable intermediate stage exists between the simplest reflex pathway and the most complicated neural pathway. It is possible to view even the most complex behavior as the result of an intricate interaction among many enormously complex reflexes."[5]

Suppose we were to climb even higher up the evolutionary scale, so high that we reached Homo sapiens. Shouldn't these same biological principles apply to the human animal as well?

Well, science does apply these same principles to humans and has noted quite a few universal behavioral patterns (in the case of humans, what are referred to as *cross-cultural* behavioral patterns) in our species, behaviors that have been exhibited in some way by every culture from the beginning of our species.

> "The essential unanimity with which the universal cultural pattern is accepted...suggests that it is not a mere artifact of classificatory ingenuity but rests upon some substantial foundation. This basis cannot be sought in history, in geography, or race, or any other factor since the universal pattern links all known cultures, simple and complex, ancient and modern. It can only be sought, therefore, in the fundamental biological nature of man and in the universal conditions of human existence."[6]

If all planarians orient themselves towards light, it implies that they are genetically programmed to do so. If all gorillas engage in particular courtship practices, this, too, would imply that they must be genetically programmed to behave in such a way.

Whether we like to believe it or not, humans are animals too. Therefore, whatever logic applies to all the Earth's other creatures must also apply to our own. If there is any behavior that has been universally exhibited among every human culture, it would imply that such behavior must represent an inherent characteristic of our species.

Take the fact that all humans from all cultures express the emotions of grief, fear, aggression, and amusement with the same exact facial expressions. All humans, for instance, express amusement with a facial expression we refer to as a smile. That all humans express amusement [smile] in the same exact way implies that, just as all planarians turn towards the light, all humans express their emotional states as the result of an involuntary, genetically inherited reflex. With this in mind, let us investigate some other, more complex cross-cultural behavioral patterns evident in the human animal.

There are many behaviors that have been labelled as cross-cultural behavioral patterns, behaviors that have been found to exist among every human society since the dawn of our species. Some examples of such cross-cultural patterns include the arrangement of kin-groups, the application of sexual restrictions, birth rites, puberty rites, marriage and death rites, acts of celebration, mourning, and courtship, incest taboos, inheritance rules, weaning, education of young, hygiene, obstetrics, status differentiation, division and co-operation of labor, community organization, development of legal codes and penal sanctions, tool-making, trade, cooking, gift-giving, joking, use of personal names, sports, dancing, singing, religious worship, music and creation of musical instruments, bodily adornment, use of calendars, counting, cosmology, belief in magic, folklore, belief in and propitiation of supernatural beings, medicine, mythology, government, and language.

So what can we say about such behaviors? As is true for all of the Earth's other creatures, whenever there is a behavior that is exhibited by every individual (again, in the case of humans, within every culture), we can presume that such behaviors must represent an inherent characteristic of that species, that is, a genetically inherited reflex or instinct.

Would this then mean that our species is genetically predisposed to engaging in something as seemingly abstract as the division of labor? Why not? Don't we see evidence of such preprogrammed behaviors among colonies of ants or bees? If such behavior can exist as a reflex in them, why can't the same be true for our species?

What about those functions that are unique to our species such as the application of math, language, music, or even religion? Is it possible that such behaviors could exist as the consequence of a reflex, a genetically inherited impulse or instinct? Let's examine one of them and see. Let's examine one that we can all accept is not only unique to our species, but exists among every human culture. Let's take language, for example.

It's accepted that all human cultures possess the capacity to communicate through a spoken language. Because we all possess this capacity, we can assume it represents a genetically inherited characteristic of our species. In addition, this would further imply that there must exist some very specific physiological site in us from which our language capacities are generated.

So where does our capacity for language originate? Does it stem from our hearts, our kidneys, our livers? Of course not. Like all cognitive capacities, ours for language originates from within the brain. How do we know this? We know this because there is physical evidence to prove it.

Within the human brain (and only the human brain*) there exist very specific structures responsible for the generation of our language capacities. Such "language" parts of the brain include Broca's area, Wernicke's area, and the angular gyrus. The angular gyrus is the part of our brain that receives sensory information such as the scent of a flower, the sight of a cat, the taste of a lemon, or the sound of a bell, and then links this sensory input to its verbal correlate or "word." For example, when we hear the sound of a bell, our angular gyrus recalls the specific word with which we were taught to define that sound. The angular gyrus therefore acts as our brain's linguistic filing cabinet, that place where all the words we've been taught to define our sense-experiences are stored. Next, Wernicke's area, which is located in the brain's temporal lobe and plays an essential role in linguistic comprehension, retrieves the recalled word from the angular gyrus and then processes it such that we can grasp the word's meaning.

---

*Though only humans are known to possess these language specific sites in the brain, there is evidence that chimpanzees can acquire certain limited language capabilities. This was made apparent in studies (Gardner and Gardner, 1969; 1971; 1978) in which chimpanzees were taught to communicate with humans through the use of ASL (American Sign Language). As some of our closest ancestors on the evolutionary ladder, it makes perfect sense that these animals should possess such incipient language capacities.

Last is Broca's area which connects the information acquired from the Wernicke's area to the part of the brain that controls the muscles of the face, jaw, palate, and larynx, thus allowing for our words to be physically spoken.

This may all sound great on paper, but how do we know such organs exist in us? We know this because in cases where any one of these sites has incurred physical damage, it has been shown to have a direct effect on some very specific part of that person's language abilities. Such linguistic malfunctions are known as aphasias. Damage to Wernicke's area, for example, which is vital to comprehension, can prevent a person from understanding words spoken to them. In some cases, a person may be able to grasp the meaning of written words that he or she cannot comprehend when heard. In other instances, damage to Wernicke's area can produce speech that, though it may be fluent, will be meaningless. Damage to Broca's area, which controls articulation, will cause impairment of speech so that articulation may be slowed, labored, or completely disabled, depending on the extent of the injury. In some cases, a person may be able to say the word "hopper," for instance, but not the word "hop." As we can see, depending on which specific parts of our language centers are damaged will determine the specific language dysfunction a person will incur.

What all this demonstrates is that there exist very specific physiological sites within our brains that are responsible for our specific language and speech capacities. No less than we all possess a heart, we all possess an angular gyrus. And again, how do such physiological sites emerge in us? From information stored within our genes. Just as we possess genes that instruct our bodies to develop a nose on our face, we possess genes that instruct our bodies to develop an angular gyrus within our brain. Apparently, there exists a specific set of "language" genes in our chromosomes from which these language enabling sites emerge. Just as our capacity to speak and comprehend a language was passed on to us through our parents' genes, we will one day pass this same capacity on to our own offspring. It is through our genes that all intrinsic human traits are transferred from one generation to the next. In other words, cognitive traits are no different from all other physical traits. The point I am once again trying to emphasize is that, even in the case of humans, universal [cross-cultural] behaviors represent genetically inherited traits.

Just as such basic physical attributes as eye color are predetermined by genetic inheritance, the same is true for our language capacities. Furthermore, just as our cross-cultural capacity to speak a language is genetically conceived, the same, we can assume, is likely to be true of all of our cross-cultural propensities.

How about music as yet another example of a cross-cultural behavior in our species? No plant, insect, fish, cat, dog, or even chimp bangs on objects to create rhythmic combinations of sound. Humans, however, do. As a matter of fact, every human culture that has ever existed has demonstrated some type of musical capacity. Does this then mean that something that we perceive as an act of inspiration such as musical creation might exist merely as the effect of a genetically inherited reflex? Is it possible that Mozart's talent may have represented the physical consequence of his being born with enhanced "musical" genes? Perhaps, for if music does indeed represent a cross-cultural characteristic of our species, it would suggest that there must exist some "musical" part of the brain from which the capacity is generated. And what evidence might there be to support such a notion? According to the musicologist, John Blacking:

> "There is so much music in the world that it is reasonable
> to suppose that music, like language and possibly religion,
> is a species-specific trait of man. Essential physiological
> and cognitive processes that generate musical composition
> and performance may even be genetically inherited and
> therefore present in almost every human being."[7]

It is generally agreed that every human culture from the beginning of our species has generated some form of music. "No culture so far discovered lacks music."[8] This would imply that if I started clapping my hands in a rhythmical manner in front of any individual from any culture, there exists the distinct possibility that he or she would have the inclination as well as the ability to join in with me. As we know, this is not something I could achieve with a plant, insect, fish, cat, or any other animal. Expressing oneself musically is, therefore, an exclusively human capacity.

In addition to the fact that music has emerged from every culture, what other evidence is there to support this notion that we might possess a "musical" part of our brain?

Let's take the capacity known as perfect pitch. This is an aptitude some people possess with which they can determine the exact pitch of any sound they hear. But perfect pitch cannot be learned. One must be born with it. This implies that the capacity for perfect pitch is innate.

What about musical "idiot savants," people born with some incredible musical ability who are intellectually retarded in almost every other way, people, for example, who after hearing a complete Beethoven sonata for the first time, can sit down at a piano and play the same piece, note for note and in perfect time, but meanwhile who can't tie their own shoelaces. We hold musical talent in such high esteem, as a trademark of human genius and inspiration. In light of such the "idiot savant," however, is this the act of an inspired genius or something more mechanical in nature, perhaps the consequence of a genetically inherited instinct, a sophisticated reflex?

How about the fact that people can suffer from musical aphasias? Similar to a linguistic aphasia, musical aphasias constitute the loss of some specific musical ability caused by physical damage incurred to one's brain. For example, after suffering a stroke, a composer may lose his ability to write music; a musician, his ability to play an instrument. That such aphasias exist suggests that our musical abilities must be integrally linked to our neurophysiological make-ups.

How about that music can affect us physiologically? "Music can provoke intense, genuine, emotional arousal from ecstatic happiness to a flood of tears."[9]

Equally revealing is the fact that, regardless of one's cultural origins, all peoples tend to interpret certain musical themes in the same way. Who, for instance, and from what culture, would ever describe a John Philip Sousa march as soothing or tranquil, as opposed to militant, triumphant, or exhilarating? Wouldn't the fact that people from different cultures experience and interpret the same musical stimuli in similar ways suggest that musical consciousness must represent an inherent part of our species' physiology?

Another phenomenon which may show that our musical capacities are physiologically based is the fact that certain combinations of sounds have been shown to trigger epileptic seizures. "Musical epilepsy convincingly demonstrates that music has a direct effect upon the brain."[10]

Without getting any more deeply involved in an argument supporting the existence of a "musical" part of the brain, it seems there exists adequate evidence to suggest that our capacity for music, like language, is directly linked to our cerebral physiologies. What this means is that music and language represent two of the means by which humans process information, two inherent modes of perception, two of the ways our cerebral make-ups determine the manner in which our species comprehends and therefore interprets reality.

After having acquired what I felt constituted adequate evidence that all cross-cultural behaviors represent the consequence of genetically inherited impulses, a complex series of reflexes, it was now time to apply this same principle to humankind's cross-cultural propensity to perceive a spiritual reality. Just as every culture from the dawn of our species has perceived the world both musically and linguistically, every culture has perceived the world "spiritually."

Was it therefore possible that humans may actually inherit their cross-cultural inclinations to perceive a spiritual reality? Were our cross-cultural beliefs in such universal concepts as a god, a soul, and an afterlife the consequence of a genetically inherited instinct, a reflex? Furthermore, if we possess such a reflex, mustn't it emerge from some specific physiological site in us, what we could perhaps call a spiritual or "God" part of our brain?

BOOK II

INTRO TO BIO-THEOLOGY

"The heart has its reasons, which reason does not know. We feel it in a thousand things. I say that the heart naturally loves the universal being."   — Pascal

"It seems that the existence of God is self-evident. Those things are said to be self-evident to us the knowledge of which is naturally implanted in us."   — Thomas Aquinas

"The predisposition to religious belief is the most complex and powerful force in the human mind and in all probability an ineradicable part of human nature."   — E.O.Wilson

ONE

THE SPIRITUAL FUNCTION

"All the civilizations of mankind that have existed were rooted in religion and a quest for God."[11]   - Ivar Lissner

Every generation of every human culture - no matter how isolated - has possessed the capacity to speak and comprehend a language. This suggests that within our chromosomes there must exist genes from which our linguistic capacities emerge in us. As we develop within our mothers' wombs, it is the role of these "language" genes to instruct our emerging bodies to forge specialized neurological connections in our brains that will eventually come to represent those language sites from which we will obtain our linguistic capacities.

What if we were to apply these same principles to human spirituality? Just as all human cultures have demonstrated a capacity for language, all human cultures have just as clearly demonstrated a distinct propensity to perceive the existence of a spiritual reality, one that transcends the limitations of this finite, material one.

"Religious belief is one of the universals of human behavior, taking recognizable form in every society from hunter gatherer bands to socialist republics. Its rudiments go back, at least, to the bone altars and funerary rites of Neanderthal man."[13]

This proclivity of ours to perceive a spiritual reality is made evident by specific spiritual beliefs and practices that have been exhibited cross-culturally since the dawn of our species. For example, every human culture has practiced a belief in a supreme spiritual entity or group of entities we refer to as a god or gods.

"There is not a civilization known to us that did not have faith in God or Gods."[12]

That every culture has maintained a belief in some form of a supreme spiritual being implies that every human culture has believed in the existence of a spiritual reality.

If it's true that all cross-cultural behaviors constitute inherited traits, then shouldn't we presume the same must hold true for our species' cross-cultural belief in a spiritual reality? Wouldn't the fact that all human cultures, no matter how isolated, have believed in some form of a spiritual force or entity imply that such a perception must constitute an inherent characteristic of our species, that is, a genetically inherited trait? Why should we view our cross-cultural spiritual proclivities any differently than we would any other cross-cultural behavioral pattern? Moreover, if we inherit our spiritual perceptions, wouldn't this further imply that we must possess genes through which this instinct is passed from one generation to the next? Furthermore, if we inherit our propensity to perceive a spiritual reality, mustn't there exist some physiological site in us from which such "spiritual" perceptions are generated? Since all perceptions are generated from within the brain, our spiritual perceptions must be generated from some portion of the brain as well. If believing in a spiritual reality represents a cross-cultural characteristic of our species, this would imply that there must exist some spiritual or, what I am informally referring to as, a "God" part of our brain.

– JUNG –

As I began to explore the possibility that we may inherit our spiritual propensities, I found there were others who had already made similar inquiries, others from whose work and research I could now borrow. Of those who had conducted such studies, it was the work of the analytic psychologist, Carl Jung, I found most pertinent. Of all Jung's contributions, however, it was his theory of the "collective unconscious" I found most applicable.

Jung's mentor, Sigmund Freud, had introduced the concept of a personal conscious and unconscious to the world. According to Freud, the personal conscious represented those thoughts, feelings, memories, and desires of which we are consciously aware. Beneath the personal conscious lay an even deeper layer of consciousness represented by an individual's unconscious self. According to Freud, one's primal drives or instincts, their personality components, memories of early childhood experiences, repressed memories and other inner conflicts all resided within one's personal unconscious. Though we might not be aware that these elements exist in us, they, nonetheless, play a significant role in all that we do, say, and think. To Freud, the personal conscious and unconscious represented the two chief components underlying all human behavior.

Jung picked up where Freud left off (something for which Freud never forgave him) by suggesting there existed an even deeper and more profound layer of human consciousness than that of our personal unconscious. Jung maintained that underlying the personal unconscious and acting as its foundation there existed what he referred to as our *collective unconscious.*

According to Jung, whereas the personal conscious and personal unconscious are derived from one's personal experiences, the collective unconscious represents those components, awarenesses, and drives we inherit and which therefore constitute an integral part of the conscious experience of every member of our species. Whereas the contents of one's personal unconscious emerge from one's personal experience and development, the content's of one's collective unconscious constitute that part of us which was forged during the our species' development and which is, therefore, common to all humankind. The collective unconscious therefore exists as a part of our inherent natures and "has contents that are, more or less, the same everywhere and in all individuals. It is, in other words, identical in all men and constitutes a common psychic substrate of a suprapersonal nature which is present in every one of us...and which have existed since the remotest times."[14]

Again, whereas the philosopher, John Locke believed we are born as a "tabula rasa," or a clean slate, waiting for our experiences alone to shape and define us, Jung, in accord with Kant, held that we are born with a set of preprogrammed cognitions and impulses. Like Kant, Jung seemed to be directing his search for answers inwards, into the nature of human consciousness.

Jung reached many of his conclusions based on his study of world mythology. Such mythologies constituted a compilation of stories, fables, legends, and parables that he found to exist among every human culture from the dawn of our species.

Through its mythology, every human culture has codified its social and spiritual norms, rites, customs, ethical standards, and beliefs. Jung not only concluded that all cultures possessed a mythology, but that all of them also contained remarkable similarities. Whether he was studying the Old and New Testaments of Judeo-Christianity, the Zarathustrian Avestas, the Norse Eddas, the Icelandic Sagas, the Islamic Koran, the Egyptian or Tibetan Books of the Dead, Hesiod's Theogony, Homer's Iliad and Odyssey, Virgil's Aeneid, the Celtic Sagas, Urartian (Armenian) cuneiform, the Japanese Kojiki (Record of Ancient Masters) or Nihongi (the Chronicles), the Babylonian tales, the Ugaritic myths of Palestine and Syria, the Chinese Shi Ching (Book of History), the Hindu Rig Veda, Mahabharata and Ramayana, the Theravada Buddhist Vinanatthu, the myths from the various cultures of Africa, Polynesia, or South and Central America, or the manuscripts of the medieval Alchemists, Jung found common themes in each of these culture's writings.

Because he found such universal similarities in the myths of every culture, Jung concluded that the contents of these myths must be generated from some inherent psychic substrate shared by our entire species. This he called our collective unconscious. Apparently, our species possessed some impulse that not only prompted each culture to create its own mythology but that fashioned each with the same universal themes. Jung referred to these common themes as archetypes. Due to the universal nature of these archetypes, Jung hypothesized that our species must possess a religious function.

> "Through the study of the archetypes of the collective unconscious we find that man possesses a religious function and that this influences him in a way as powerfully as do the instincts of sexuality and aggression. Primitive man is as occupied with the expression of this function, the forming of symbols, and the building up of religion as he is with tilling the earth, hunting, fishing, and the fulfillment of his other basic needs."[15]

Inspired by Jung's theories, particularly by his suggestion that humans possess a "religious function," I now felt, more than ever, that human beings might indeed inherit their spiritual sensibilities. The chief difference between my interpretation and Jung's was that, whereas Jung perceived human consciousness as a function of the human mind, I saw it as a function of the brain. Once again, whereas the mind implied the possibility that there might exist some intangible/transcendental component in us, the brain did not. Whereas advocates of the mind might perceive cognition to be a function of the soul, advocates of the brain perceived cognition to be a function of one's neurophysiology.

Based on what the bio-psychological (or psycho-biological) sciences had taught me (sciences that simply weren't available to Jung in his time), I had adopted a strictly physical/organic/physiological/mechanistic, that is, scientific approach to human sensation, perception, emotion, and cognition, that is, to human consciousness.

What if I were to apply these newly advanced bio-psychological sciences, not just to consciousness, but, more specifically, to Jung's notion of the collective unconscious? What if it were possible to "biologize" the collective unconscious? What if what Jung spoke of as a religious function could be explained as a genetically inherited trait? By applying the bio-psychological sciences to the study of human spirituality, I now felt it might be possible to construct a purely physical/organic/mechanistic, that is, a scientific interpretation of human spirituality as well as of God.

– UNIVERSAL SPIRITUAL BELIEFS AND PRACTICES –

As I explored the world's various religions and mythologies, it became apparent that every human culture from the dawn of our species has maintained a dualistic interpretation of reality. In other words, every human culture - no matter how isolated - has perceived reality as being comprised of two distinct realms, the material [physical] and the spiritual. According to this cross-cultural perception, each realm is composed of its own distinct substance, the material realm being composed of matter and the spiritual being composed of spirit - a concept for which every known culture has either possessed a symbol or word.

All things that exist as a part of the material realm, that is, all things composed of matter, are cross-culturally perceived as existing in a state of constant flux, susceptible to the forces of birth, change, death, and decay, rendering them destructible, fleeting, finite. They are corporeal, capable of being perceived with our physical senses and therefore empirically verifiable.

Things that exist as a part of the spiritual realm are cross-culturally perceived as transcending the physical forces of nature and therefore as being indestructible, infinite, and everlasting. Because things that exist as part of the spiritual realm are perceived as being incorporeal, they are incapable of being detected by our physical senses.

This universal perception that there exists a spiritual reality is made evident by a number of cross-cultural beliefs and practices. For instance, every culture has universally expressed a belief in the concepts of a spiritual realm, a supreme spiritual being [a god or gods], a soul, and an afterlife. The universality of these beliefs are made evident by a series of universally performed rites and practices.

For instance, every culture has exhibited a tendency to pray to, worship, and petition a god or gods. This is made evident by the fact that every culture has erected sites of worship from which members of its community can conduct their prayers. Whether it be a Muslim mosque; a Catholic church; a Jewish synagogue; an ancient Aztec, Greek, or Egyptian temple; a Shinto shrine; a Babylonian ziggurat; or the underground ceremonial chamber of the Anasazi, every culture has constructed physical edifices specifically designed for the purpose of praying to and petitioning one's gods. Such sites of worship constitute physical evidence that all cultures have believed in the existence of a spiritual reality.

In addition, every culture has created religious works of art. The first examples of this exist in the form of cave art which dates as far back as to the early paleolithic age, from about 40,000-12,000 BC. Such cave paintings often depicted representations of a hunt in which an animal is covered with javelin wounds highlighted with red ochre. Because the spear designs were often painted over one another, it is believed that these paintings were constantly renewed for magico-religious purposes to effect a kill in the chase.

In spoken and written form, every culture has expounded upon its spiritual beliefs through scriptures and mythologies. That all cultures have possessed such tangible works of art and text constitutes further evidence that humans cross-culturally believe in a spiritual reality.

Furthermore, every world culture has maintained a belief that humans possess a spiritual component within ourselves. This spiritual component, as it exists within us, we refer to as a soul - another concept for which every culture has possessed either a symbol or word. "The soul is a universal concept."[16]

According to our cross-cultural belief in a soul, humans perceive themselves as being comprised of a unique combination of both matter and spirit. While we cross-culturally perceive our bodies to be constituted in matter, we, at the same time, cross-culturally perceive human consciousness as being constituted in spirit, what we refer to as a soul. In this way, we project our dualistic conception of reality onto our own selves.

Just as we perceive things that consist of spirit as being indestructible, eternal, and everlasting, we perceive our souls as possessing these same attributes. In other words, we believe that by virtue of our souls, we, our conscious selves, are indestructable, eternal, and everlasting. This leads us to cross-culturally believe that though our physical bodies will perish, our spiritual selves, our spirits or souls, will continue to endure for all eternity. It is through this universal belief in a soul that human beings derive their sense of immortality.

"Through religion man affirms his convictions that death is not real nor yet final and that we are endowed with a personality which persists even after death."[17]

This notion that all human cultures have believed in a soul as well as in their own immortalities is supported by the fact that all human cultures have expressed a belief in an afterlife, "a new or continued or transformed existence after death, belief in which has been found in virtually all cultures and civilizations."[18] Whether we are speaking of Heaven, Purgatory, Hell, Valhalla, Niflheim, Nirvana, Tartarus, the Elysian Fields, Hades, Oblivion, the Realm of the Dead, the spirit land (Te Reinga), the Mystical Garden, Paradise, reincarnation, or transmigration of the spirit, all cultures, both eastern and western, have expressed a belief that our spiritual selves or souls persevere long after our physical bodies have perished.

This universal belief in an afterlife has been physically manifest through the cross-cultural practice of a funerary or burial rite. Not only have all cultures practiced the disposal (generally burial) of their dead, but, most significantly, this act is cross-culturally performed with a rite that anticipates sending that individual's spiritual component or soul onwards to some next or other realm. As further evidence, many cultures have buried their dead with artifacts meant to facilitate the deceased person's transition from this realm to the next. In essence, every culture has believed there exists some transcendental quality in us that endures after our physical bodies have perished.

Burial represents the last of a series of cross-culturally practiced rites through which the individual's soul is sanctified before its culture's gods. While burial represents the last of these rites, all cultures inaugurate the newly born into their spiritual community with a rite of birth. Whether it be a Jewish or Muslim circumcision, the immersion of a Catholic child into the baptismal font, or the Australian Aborigine rocking its newborn through the purifying smoke of the Konkerberry fire, every culture sanctifies its newly born with a rite in which the young are introduced into the spiritual community. As the cultural anthropologist, Mircea Eliade, expressed in his book, *The Sacred and the Profane*, "When a child is born, he has only a physical existence; he is not yet recognized by his family nor accepted by the community. It is only by virtue of those rites performed immediately after birth that he is incorporated into the community of the living."

As the child grows, he or she goes through a series of life passages. After the birth rite, the next passage to be cross-culturally recognized in a spiritual format is exhibited in the form of an initiation rite. This rite, which is usually celebrated in tandem with puberty, signifies one's passage from childhood to adult, and is meant to sanctify an individual before his gods as a grown and responsible member of the spiritual community. Whether it be a Jewish Bar-Mitzvah, a Congolese Kota face-painting ceremony, a Catholic Confirmation, an adolescent Baptism of the Southern Baptist, or a Hindu Sannya ceremony, every culture ritualistically assimilates its individuals into the spiritual community as an adult. Using Jungian terms to express the cross-cultural nature of this rite, the author, Anthony Stevens, writes in his book, *On Jung*, "Comparison of rites from all over the world suggest that these initiation rites themselves possess an archetypal structure, for the same underlying patterns and procedures are universally apparent."

After being initiated into the spiritual community, members of the opposite sex are united in order to promote procreation. Such unions are consecrated through a cross-culturally practiced marriage rite.

More evidence that points to our species' cross-cultural belief in a spiritual realm rests in the fact that each of the aforementioned rites are presided over by some member of the community who is assigned the role of spiritual guide or leader. Every culture has possessed some form of a priesthood, some individual or group of individuals whose role is to act as the community's intermediary between the material and spiritual worlds. Whether this individual is referred to as a shaman, priest, rabbi, swami, yogi, oracle, mystic, psychic, medium, or imam, all cultures have possessed some such member, group, or caste whose role is to serve the community's spiritual needs.

To further confirm our species' cross-cultural nature to believe in a spiritual reality, all cultures have demonstrated a tendency to ascribe magical, sacred, or supernatural, that is, spiritual properties to certain people, places, or objects.

As mentioned, all cultures have possessed sites of worship. In addition to these, every culture has also ascribed "sacred" status to a number of sites referred to as shrines. Whether it be the Tomb of the Patriarchs, the Kaaba Stone, Delphi, the Pyramids, the Dakhma of Cain, the Ganges River, Bethlehem, or a Buddhist Stupa, each represent centers of pilgrimages and adoration because of their spiritual significance and the spiritual values they've come to symbolize.

Sacred status has also been cross-culturally ascribed to various objects. Totems, relics, icons, amulets, talismans, charms, or fetishes, as they are called by their various cultures, represent examples of physical objects believed to carry the substance of spirit within them. Whether it be the wafer and wine of the Eucharist, the ceremonial Calumet or Peace Pipe of the Native Americans, the hairs of the prophet Mohammed, the sacred tooth of Buddha, fragments of the holy crucifix, a mezuzah, an African gris-gris, or a crystal for the new-age spiritualist, all represent material objects that are believed to possess a magical or "spiritual" quality. That all cultures have assigned sacred status to physical objects further attests to the fact that all human cultures have believed in the existence of a spiritual reality.

Furthermore, all cultures have also expressed a belief in the existence of spiritual/transcendental forces that guide and influence all that transpires in the world. This is made evident by our cross-cultural beliefs in such abstractions as luck, karma, kismet, fate, fortune, and destiny. Such concepts demonstrate our cross-cultural perception that there exist transcendental forces which influence and intervene in all that occurs within the material universe.

Humankind's universal belief in a spiritual element is further evinced by the fact that all cultures tend to associate the sentiment of guilt in a religious context. Though we may feel guilty for things we've done to other men, all cultures show an express concern for how they will be judged for their actions by their gods. This is made evident by a variety of rites of atonement and penitence through which individuals from every culture have sought to repent for those crimes committed against their gods, otherwise known as "sins," yet another concept for which every culture has possessed a word. Physical evidence of penitent behavior is manifest by a variety of sacrificial rites. In such rites, individuals or the group as a whole makes offerings to the gods in the hope that it will solicit their sympathy, mercy, or forgiveness. We seek divine mercy and forgiveness because we believe that our actions can influence the quality of our lives both here on Earth as well as in the afterworld. The universality with which the sentiment of guilt seems to be exhibited in a religious context suggests that it may somehow be associated with our inherent spiritual sensibilities.

So, all human cultures have practiced a belief in the existence of a spiritual realm, a God or gods, a soul, and an afterlife. Strange that we should all perceive reality with this same spiritual bent, that we should all hold such similar beliefs and then express them through such similar rites and practices. Is it possible that we all practice such similar beliefs and then express them through such similar rites as the result of some vast coincidence? Or is this the result of something deeper, something more essential to our inherent neurophysiological make-ups at work here?

Similar to the manner in which all planarians have a tendency to orient themselves towards the light, the fact that all human cultures have a tendency to believe in a spiritual reality would imply one of three things. The first possible reason that all cultures have conceived of the same spiritual concepts would be as the result of some vast coincidence. This would be tantamount to believing that all planarians orient themselves in the direction of light for the same reason. Both possibilities, I imagine, are equally unlikely.

The second possible reason that every culture has shared the same beliefs and then practiced them in such similar fashions is that during the emergence of our species the concepts of a spiritual realm, a God, a soul, and an afterlife were created by a few inspired individuals whose innovative ideas were verbally passed from one generation to the next as our species spread across the continents. This would imply that our cross-cultural belief in a spiritual reality represents a learned as opposed to an inherited behavior.

The problem with this possibility is that, as our species spread across the globe, it's hard to imagine that any one set of learned beliefs and behaviors could have so persistently endured. Learned, as opposed to inherited behaviors, come and go like the wind. This is why, for instance, though a variety of words have flourished and then become obsolete throughout the history of our species, language, in general, has existed among our species as a constant. The same, I am suggesting, is true for our spiritual beliefs. Though scores of religions have flourished and disappeared throughout the history of our species, religion, in general, has existed as a constant. Similarly, though scores of various rites, practices, and beliefs have come and gone with time, the more fundamental beliefs in a spiritual realm, a God or gods, a soul, and an afterlife have persisted throughout. It is these fundamental beliefs that represent the foundation of every world religion. It is simply the manner in which these primary beliefs have been manifest that is constantly shifting and evolving. The fact that these primary beliefs have so tenaciously persisted in every culture, and under such diverse environmental and historical circumstances, leads me to believe that, just as in the case with language, there must be some underlying physiological force at work here.

Take, for example, our feelings of grief or sadness. Why is it that all humans express these sentiments in the same way? Why is it that all humans cry? No one has to be taught to shed tears when mourning the death of a loved one. This is something we do innately, a reflex, a specific response to a particular stimulus. But let's imagine for the moment that crying was a learned behavior. Let's imagine that we had to be taught to cry as a means of expressing our sadness. If this were the case, wouldn't it be likely that at some point in time, some culture - just one - would have deviated from its original teaching and eventually come up with some other means of expressing this emotion? If crying were a learned behavior, it is incredibly unlikely that every culture on earth would, to this day, all express grief in the same exact way. Analogously, I propose that the same principle can be applied to our universal spiritual beliefs and practices.

If such concepts as a belief in a spiritual realm, a God, a soul, and an afterlife were merely invented by a few inspired individuals during the dawn of our species, it seems highly unlikely that every world culture would still be practicing these same beliefs today, and all in such similar ways.

This leaves us with the last possibility which asserts that our universal spiritual beliefs must represent an inherent characteristic of our species, a genetically inherited trait. This would suggest that we inherit our spiritual proclivities and are innately predisposed to believing in the existence of a spiritual reality. If true, then such an inherent trait could only come from one place: information stored within our genes - thus implying that humans possess "spiritual" genes. This further suggests that we must possess a physiological site or group of sites, a specific group of neural clusters from which our spiritual perceptions, cognitions, sensations, and impulses are generated.

Imagine we were to study ten separate and totally isolated colonies of honeybees, all of which constructed their hives in the same hexagonal design. After witnessing this, would we say that such behavior represents an example of "free thinking" bees all coincidentally building their hives in the exact same way? Or would we, instead, say that the bees, as a species, must be physiologically "hardwired" to construct their hives in such a way, that is, that they do this as a reflex, an instinct? Do all bees construct their hives in the same hexagonal fashion because they "chose" this design or because they are preprogrammed to do so? In such a case, I imagine we would agree that the fact that all bees construct their hives in such a similar fashion must be as the result of a physiological impulse, that somewhere in the bee's brains there must exist a cluster of neurons that compel them to construct their shelters this way.

With this in mind, why, I ask, should we view our own universal [cross-cultural] behaviors any differently than we might the behavior of bees? In the words of the pioneer (if not the founder) of the science of sociobiology, E.O. Wilson, "The same principles of population biology and comparative zoology that have worked so well in explaining the rigid systems of the social insects could be applied point by point to vertebrate animals." If we are ever to make any progress in the understanding of our own physical natures, mustn't we study and assess ourselves with the same objectivity that we do all the Earth's other creatures? If a group of aliens were to study our species from above, what might they conjecture after witnessing approximately one hundred thousand years of the vast majority of our species disposing of its dead in a hole in the ground performed with a particular rite? Would they not view such behavior as representative of an instinct? Would they not regard our behavior similar to the way we view the universality with which all planarians turn towards the light or bees construct their hives? Wouldn't these aliens surmise that the burial of the dead must represent an inherent characteristic of our species, the behavioral consequence of a genetically inherited instinct?

In the same way that planarians are "wired" to turn towards the light, humankind is "wired" to believe in a spiritual reality. Because this impulse is cognitive in nature, it must originate from a part or parts of the brain. In other words, there must exist particular clusters of neurons within the brain from which our spiritual cognitions, perceptions, sensations, and impulses are generated. This would further suggest that should we alter these clusters of neurons, these spiritual parts of our brain, it would affect one's spiritual consciousness.

Should this part or parts of a person's brain, this spiritual function, be surgically removed, for instance, it's highly likely that that individual would lose his sense of spiritual consciousness. Never again would that person have a spiritual experience. Never again would he feel the warm presence of some protective spiritual force or entity. Never again would he feel compelled to pray, to look outwards to some transcendental being for guidance. Similar to the manner in which a person can develop a linguistic aphasia, I'm suggesting that it's possible to develop a spiritual one. For example, when a priest suffers from Alzheimer's disease, does he not lose, along with his other sensibilities, his sense of spiritual consciousness? Are we to believe that though this person can't feed or go to the bathroom by himself, he will still be able to pray or preach the gospel with lucidity? Apparently, spiritual consciousness is just as integrally linked to our neurophysiological make-ups as is any other of our other cognitive capacities.

Perhaps most controversial of all, if what I'm suggesting is true, it would imply that God doesn't exist as something "out there," beyond and independent of us, but rather as the product of an inherited perception, the manifestation of a biologically-based evolutionary adaptation that exists exclusively within the human brain. If such a hypothesis is correct, it would imply that there is no spiritual reality, no God or gods, no soul, and no afterlife. Such spiritual concepts as these would only exist as manifestations of the particular way that our species has been "wired" to perceive reality. In such a light, humankind can no longer be viewed as a product of God, but rather, God must be viewed as a product of human evolution, the perceptual manifestation of our species' inherent cognitive processing.

Just as Kant had proposed that we inherit our sense of temporal and spatial consciousness, I'm suggesting we inherit our sense of spiritual consciousness. Furthermore, just as Kant suggested that we are born with spatial and temporal modes of perception, two means through which our species is "wired" to interpret reality, I'm suggesting that spirituality represents yet another one of these inherent modes of perception, another way our species is genetically programmed to interpret reality. This would imply that our spiritual perceptions, like all others, are not representative of any absolute truth, but are merely relative to the manner in which our species is programmed to interpret reality.

Not only does this spiritual function act to transform our perception of reality, but it also seems to possess the ability to override our empirical/analytical/logical awarenesses. This is made evident by the fact that regardless of how little physical evidence there is to support the existence of a spiritual realm, every culture has believed in one. This is rather unusual for a species of skeptics as ourselves. Generally speaking, human beings tend to believe only what their physical senses reveal to them. Unless we can see, feel, taste, smell, or hear something, we tend to be dubious of its existence. Even so, our spiritual beliefs seem to represent an exception to this rule. Since there is no physical evidence to support a belief in any spiritual reality, our spiritual beliefs and perceptions must be originating from somewhere within our own heads rather than from information acquired through our physical senses.

For instance, should I tell the *average* person from any world culture that there were invisible pink elephants hovering around the room, chances are he would shun me, if not try to medicate me. He would react in such a way simply because the information I'd be conveying would contradict everything his physical senses would be revealing to him. Nevertheless, should I tell the same person that the spirit of God or the deceased were hovering about the room, chances are he would be much more inclined to believe me, regardless of what information his physical senses might convey. That we possess a cross-cultural tendency to perceive reality with such a spiritual bias would suggest that we are physiologically compelled to do so.

To reiterate, if we apply the principle that all universal [cross-cultural] behaviors represent the effects of genetically inherited reflexes, it would suggest that human beings are "wired" to believe in the concepts of a spiritual reality, a God or gods, a soul, and an afterlife; to pray and to worship; to ritualistically bury their dead; to conduct birth, initiation, and marriage rites; and to undergo "spiritual" experiences (something on which I'll elaborate in chapter three). This would further suggest that for every cross-cultural spiritual cognition, perception, or sensation we experience, there must exist some specific cluster of neural connections in the brain from which they are generated. Damage to any of these neural clusters would consequently impair whatever specific spiritual perception, sensation, or cognition that happens to be generated from that particular site. In summary, such a hypothesis would imply that all of our "spiritual" cognitions, perceptions, sensations, and behaviors are the manifestations of inherited impulses generated from neural connections in the brain and, therefore, not indicative of any actual spiritual reality*.

But why - one might justifiably ask - if all cultures are instilled with the same inherent "spiritual" impulse, do such a wide variety of religions exist?

Though we all possess the same sites in the brain through which our linguistic capacities are generated, each culture - based on its unique set of historical and environmental circumstances - develops its own linguistic identity or language. Analagously, though we all possess the same sites in the brain through which our spiritual inclinations are generated, each culture, based on its unique set of historical and environmental circumstances, develops its own spiritual identity or what we call religion. Religion therefore represents the cultural manifestation of the spiritual impulse. (As a matter of fact, it's quite possible that religiousity stands as its own distinct impulse, unique from our inherent spiritual sensibilites, one that regulates our need to engage in series of repeated rituals. As a matter of fact, it's possible that obsessive compulsive disorders represent the dysfunction of this very impulse, one meant, in it's healthiest form, to promote religious ceremony and ritual.)

Similar to the way all languages share the same fundamental rules of construction and syntax, each religion shares the same fundamental beliefs. Though each culture may believe in a different god, each believes in the existence of some supreme transcendental force or being. Though each culture may hold its own view of what death will bring, each believes in some form of an afterlife. Again, though

---

*Though no one could ever prove that there is no such thing as a spiritual reality, such a hypothesis would certainly support the possibility that one might not exist. As a matter of fact, it is impossible to prove that any imaginary force or being does *not* exist. How, for example, could one ever prove that there are no such thing as invisible dragons? Just because we've never seen one, doesn't prove they don't exist. As a mattter of fact, the mere aim of attempting to prove that an imaginary being doesn't exist is an exercise in futility. We must accept the principle that the burden of proof need lie in confirming that something does exist as opposed to in that it merely does not.

we might all possess the the same "spiritual" genes, the same "spiritual" function, because each culture has emerged from its own particular environmental and historical circumstance, each has developed its own religion.

This might, for instance, help to explain, why, whereas more northern cultures, as the Norse, have incorporated such indigenous animals as bears, wolves, and whales into their religions, desert-based peoples, as the ancient Egyptians, have incorporated animals as jackals, falcons, and snakes into theirs.

Presuming that spirituality represents the product of a genetically inherited impulse, I next had to ask why we would have evolved such a trait. What prompted the forces of evolution to select such a seemingly abstract characteristic as spirituality into our species? As all traits must serve to enhance a species' survivability, how might a spiritual function do this for ours? Furthermore, what is it about our species, in particular, that we alone should possess such a trait?*

Unless I could provide a sound explanation, a rationale, for why such a spiritual function might have evolved in us, it would be impossible to justify one's existence. So the question now was: Why would our species have evolved a spiritual function?

*With the exception of Neanderthal Man's simple bone altars and burial rites, no other species, besides our own, has given us any reason to believe that it might be spiritual in nature. Nevertheless, I have had others contend that this is a presumptuous assertion to make given that we can never really *know* what another animal is thinking. How can we *know*, for certain, that no other species senses a spiritual reality or believes in a god? Granted, though one can never get inside an animal's thoughts, based on their behavior, we have no reason to believe that any animal besides our own possesses a spiritual consciousness. When, for example, have dogs gathered around a ceremonial mound they erected and then bowed their heads in what might be suggestive of an act of deference or prayer? When has any chimpanzee carved or drawn a symbolic image of some imaginary or "spiritual" force or being? When has any other animal (besides the aforementioned Neanderthals, who were extremely close cousins of ours) buried its dead in a ritualistic manner, suggesting that it might conceive of some form of an afterlife? It is through an animal's behavior that we gain insight into the inner workings of their conscious experience, and none, other than our own, has given us any reason to believe that it possesses any semblance of spiritual consciousness.

TWO

THE RATIONALE

"All that exists is rational." - Hegel

All that exists is rational. Every cause has its effect; every effect has its cause. In essence, nothing happens without a reason. Since this axiom applies to all that exists, it must also apply to all the various life forms that exist on this planet as well - all forms, including our own. For every trait we possess, from stereoscopic vision to upright posture, from opposable thumbs to our language skills, each must have its reason for being. Since the primary impetus that drives all evolution is the preservation of a species, every trait we possess must somehow serve to increase our species' chances of survival. This is evident in every organ we possess (excluding, of course, those vestigial parts such as the coccyx [that evolutionary memento of our predecessors' tails] and the appendix [a relic of our grass eating days], two examples of parts which, because we no longer need them, are in the process of being selected out of us). Because all traits must perform a specific function that will serve to increase a species' survivability, if humans possess a spiritual cognitive function, then the same must hold true for this part as well.

With this in mind, the question becomes: if humans do possess such an adaptation as spiritual consciousness, what might its purpose be? What function might such an adaptation serve that it could enhance our species' survivability? What is such a trait's rationale, its reason for being? As is true of all traits, if human spirituality didn't provide some very specific function, if it didn't somehow enhance our species' survivability, it would never have emerged in us.

All organic attributes, that is, all physiological traits, emerge in response to some environmental pressure. For instance, if arctic wolves possess thick coats of fur, it's because their environments "pressured" them to evolve one. As our terrestrial environments are in a state of constant flux, organic matter - life - is constantly being forced to adapt to meet the demands of our ever changing conditions. Therefore, if humans possess a spiritual cognitive function, we must not just seek to understand its purpose, but we must also seek to understand those environmental pressures that may have forced the evolution of such an adaptation upon us.

In the case of wolves, it was the pressure incurred by the cold weather that caused their heavy coats of fur to emerge. What, then, was it about our environment that may have prompted the evolution of a spiritual function on our species? Furthermore, what was so unique about our species that we alone should develop such a seemingly abstract trait? Given that nature weeds out all that is superfluous, if a spiritual function didn't somehow promote our species' survivability, one simply wouldn't exist.

– AWARENESS OF DEATH –

"No thought exists in me which death has not carved with his chisel."   – Michelangelo

"In a hundred countries, in a thousand languages, humanity stops and reaches upward, keenly aware of its mortality."   – Peter Matthiessen

No other creature on earth has the intellectual capacity of Homo sapiens. As a matter of fact, our intelligence constitutes the foundation of our species' strength. Whereas fish can swim, birds can fly, and cats have speed, humans possess an intelligence that has allowed us to venture deeper, fly higher, and move faster than any other creature on Earth. No other creature (besides the quasi-living viruses) comes close to challenging our dominion over the other forms of life. All we have to do is to look around us to behold the awesome power of our intelligence. In the last hundred years alone, we have transformed this planet's surface more dramatically than any other species has in the last three billion.

But all that glitters is not gold! For as much as intelligence has graced our species, it has also been the source of our greatest affliction. Though our intelligence may have made us the most powerful creature on earth, this same adaptation has backfired on our species with nearly the same potency that it has served us. As a result of our intelligence, something happened that had never before occurred within the known universe. With the same powers of perception that had allowed our predecessors to scrutinize the world around them, Homo sapiens had developed the unique capacity to perceive their own selves. For the first time in the history of life, an organic form emerged that was aware of its own existence. No other creature before ours had any idea, for instance, that when it drank from the watering hole, the image it gazed down upon was that of its own reflection. Now, for the first time in life's three and a half billion year history, an organism - ours - suddenly could. For the first time in the history of the known universe, a combination of molecules had emerged that could comprehend the concept of its own existence.

Imagine those first primal humans looking down at their own hands, their own bodies, in awe of what they saw, and for the first time in terrestrial life's history asking that fateful question, "What is this that I am? What is this that I exist?" With the capacity for this one cognition, this one self-reflection, the human species was transformed. In biblical terms, Man had just taken his first bite from the Forbidden Fruit, fresh off the Tree of Knowledge.

It was probably not long after this first cognitive lightning flash that we were hit with the inevitable thunder: "If I am, if I exist, then isn't it possible that one day I might not?" With the same capacity with which humans could comprehend their own existence, we became equally aware of the possibility of our own non-existence...of death. With this one awareness, the wheels of life, which had been turning so smoothly for all these billions of years, had turned down a cognitive cul de sac. Humankind had suffered life's first existential crisis.

– THE PAIN FUNCTION –

According to Buddha, enlightenment could be attained by anyone who was willing to follow the path of the "Fourfold Truths." The
first of these truths, which Buddha referred to as Dukkha, asserts that
life is a process of universal misery and suffering. No matter who we
are, be it prince or pauper, we are all destined to experience the same
fateful demise. We are all bound to grow old, weak, and infirm. We
are all bound to lose everything we ever had or loved, including ourselves. In a nutshell, we are all going to die.

Borrowing from this tenet of Buddhist pessimism, Freud
expressed a similar notion:

> "We are threatened with suffering from three directions:
> from our own body, which is doomed to decay and disso-
> lution and which cannot even do without pain and anxiety
> as warning signals; from the external world, which may
> rage against us with overwhelming and merciless forces
> of destruction; and finally from our relations to other
> men."[19]

Because our lives are incessantly threatened by such perilous
forces, pain represents not only a strictly biological phenomenon but
a biological necessity. Just as with every other trait we possess, pain
exists in us because it serves a very specific function.

But what exactly is pain? Pain is a negative sensation experienced by organic forms when specific receptors are triggered by
some potentially hazardous stimulus. Stimuli that elicit pain are generally indicative of things that represent potential threats to an organism's physical existence. For example, fire or excessive heat can
harm, if not kill, a creature. It is for this reason that many animals
possess receptors that cover the surface of their skins that are heat-
sensitive. When these receptors are exposed to excessive heat, an animal experiences this potentially hazardous stimulus as the negative
sensation we call pain. By experiencing excessive heat in such a negative or "painful" manner, animals are motivated to avoid that which
can burn them. Should an animal accidentally get too close to a
flame, the negative sensation of pain will motivate it to retreat, thus
saving it from what may have caused more serious, if not irreparable,

damage. Pain therefore represents an evolutionary adaptation meant to prompt organic forms to avoid those things which represent a potential threat to their existence. It is this pain function which keeps us ever-vigilant. It is what prevents us from allowing ourselves to be cut or pierced by sharp objects, to burn, freeze, starve, or dehydrate.

To provide a more specific example of how this pain function works, I'll refer to the stimulus of hunger in a rabbit. In order to prevent a rabbit from starving to death, its undernourished body will send a distress signal to its control center, its brain, that it is in need of sustenance. It is this negative sensation that will motivate the rabbit to seek its required fuel supply. If this physical need is not met within a certain time frame, the animal's body will reinforce this signal by stimulating even more pain receptors, causing the rabbit's hunger to be intensified. What was previously experienced as a mild discomfort becomes an even more acute pain. In essence, the body is sending itself a distress signal, "Fuel me or die!" In order to relieve itself of the painful sensation of hunger, the animal is motivated to seek food and to eat.

Now let's suppose the rabbit finds itself some fuel, what we call food. In our own inaccurate language, when the rabbit finally consumes its meal, we tend to say that it is experiencing pleasure or satisfaction. But if we look at this from a purely biological standpoint, it is not pleasure that the animal is experiencing but rather the diminishment of its pain.

Just as the experience of pain works to increase an animal's survivability, it plays an equally important role in maintaining the preservation of a species. For example, it is the negative stimulus of sexual tension that incites all animals to reproduce. Among mammals, for example, reproduction represents a disadvantage to the individual, as having to provide for offspring means an animal has that much less time to devote to securing its own personal needs. Bearing children therefore represents a hindrance to individual survival. Nevertheless, as reproduction plays an integral role in the preservation of any species, it is a necessity. It is for this reason that animals are incited to engage in sexual intercourse. Once again, though sex is perceived as a pleasurable experience, it more accurately represents the diminishment of sexual tension or pain.

Among the "higher" social animals, an example of a negative or painful stimulus that serves to promote the well being of both the species as well as the individual involves the experience we refer to as loneliness. When one feels alone, that is, separated or isolated from his group or community, he is most vulnerable to life's hazards. As no individual is completely self-sufficient, each of us must rely on the assistance, care, and protection of others in a variety of ways. On our own, we are most vulnerable. Within the group, however, an individual gains the added security and strength that comes with increased numbers. It is for this reason that a negative stimulus was selected into us that prompts individuals to pursue the company of others. It is for this reason that feelings of separation and isolation elicit a discomfort we commonly refer to as loneliness, or in some cases, even depression. This particular painful sensation incites the individual to seek out and to bond with others within the group. This, in turn, increases the individual's as well as its group's chances of survival.

In summary, it is pain that keeps organic forms alive and intact. Pain is nature's electric prod that is incessantly goading us towards those things which benefit us and away from those which can harm us. We therefore experience pain and discomfort for a reason. Pain represents the chief stimulus by which all life is motivated to survive.

– THE ANXIETY FUNCTION –

Among the "higher" order animals, most particularly the mammals, threatening circumstances elicit a particular type of pain we refer to as anxiety. Anxiety constitutes a specific type of painful response meant to prompt these higher order animals to avoid potentially hazardous circumstances.

As the stomach is the organ responsible for the digestion of food, its pain receptors respond to the quality of the nourishment it receives. Analagously, as the brain is where all data is stored, it's responsive to the quality of information it receives.

For example, a baby rabbit pokes its nose into a fire for the very first time. The excessive temperature stimulates the rabbit's heat receptors dispersed throughout its skin. The negative signal stimulates the motor reflexes, which prompt the rabbit to withdraw from the flame. Having escaped the situation with no more than a superficial burn, the rabbit will now encode this painful experience in the form of a memory. From now on, whenever the rabbit perceives a fiery object, the memory of its encoded experience will be retrieved, thereby warning it not to make the same mistake. Rather than having to experience being burned over and over again, the rabbit's memory will now act as a buffer against all possible future experiences with objects that emit excessive heat.

Though this capacity to store and utilize memories enables the rabbit to avoid fire without having to be burned over and over again, this does not mean that the memory itself is altogether pain free. In order to remind the rabbit that fire is hazardous, the memory will elicit a certain amount of discomfort we call anxiety. It is necessary that this memory elicit a certain amount of discomfort (anxiety) in order to keep the rabbit ever-aware of the potential threat that fire and excessive heat represent. By virtue of its stored memories, the mere sight of fire will cause the rabbit to experience a discomfort (anxiety) that will induce it to avoid unnecessarily close contact with all future flames. In this way, though anxiety may serve to protect the rabbit from incurring actual physical injury, it, nonetheless, provides it with a certain degree of discomfort. That an actual memory can cause one to experience such psychological discomfort (anxiety) demonstrates that memories can store emotional as well as purely perceptual data. As a matter of fact, emotional memory can be attributed to the brain's amygdala, which, when damaged, will result in the loss of an individual's capacity to retrieve memories that contain emotional content (Le Doux, 1994)

With the advanced capacity to store memories, in conjunction with a capacity for anxiety, an organism no longer had to sustain actual physical injury before it was motivated to avoid a potentially hazardous circumstance. Anxiety, therefore, acts as an early warning device that keeps an organism ever alert to potential threats before one is actualized.

In another more extreme example of how anxiety serves us, imagine that the rabbit now wanders off into a cave to find itself face to face with a fierce and hungry mountain lion. The urgency of the situation causes the rabbit to experience all of the painful symptoms of anxiety, meant to motivate it to escape such a potentially hazardous circumstance. Some of the negative symptoms of anxiety include perspiration, heart palpitations, muscle tension, fear, and trembling - all of which are extremely uncomfortable and meant to incite the rabbit to get as far away from the source of its discomfort (in this case the mountain lion) as quickly and effectively as possible. Even though the mountain lion has yet to lay a paw on it, the rabbit will still experience the pain of its own anxieties which are meant to motivate it to escape the source of its threat. In the case where an animal is confronted by a mortal threat as this, the symptoms of anxiety are extremely painful. Anxiety therefore serves as an advantageous adaptation that prompts an animal to respond to a potentially hazardous situation with greater speed and efficiency. Should the rabbit manage to escape the mountain lion, it will encode this anxiety-engendering experience in the form of a memory. Now, the next time the rabbit leaves its lair, the anxiety-inducing memory of its past experience with the mountain lion will prompt the rabbit to avoid going anywhere near one. Thanks to this anxiety function, our rabbit no longer needs to be attacked by a mountain lion again to know to avoid one. In this regard, anxiety represents a biological necessity. As Ernest Becker, author of *Denial of Death,* wrote:

> "Animals, in order to survive have to be protected by fear responses, in relation not only to other animals but to nature itself. They had to see the real relationship of their limited powers to the dangerous world in which they were

immersed. Reality and fear go together naturally."[20]

Because the human brain is larger and more complex than that of all other species, our cognitive capacities are more sophisticated. First of all, our brains contain more storage space, enabling us to retain more memories. Secondly, our species possesses an enhanced capacity to consider possible future events. For example, because human beings know that the experience of hunger will elicit pain, combined with our enhanced capacity for foresight, this prompts us to concern ourselves with the procurement of food and shelter far into our futures. Unlike many of the lower-functioning animals who need to rely on the immediate stimulus of hunger to be prompted to search for nourishment, human beings are compelled to make sure there is food available long before it's actually needed. This capacity for foresight grants us the added benefit of having more time to secure our most basic physical needs. Because a simpler organism needs to rely on the immediate stimulus of hunger to be motivated to search for its food supply, it may only have a few days' advance notice to procure its next meal before it will die. In the case of human beings, however, as a result of our advanced capacity for foresight, we can erect entire structures whose sole purpose is to store food.

Though this capacity for foresight may work to our advantage, it also has drawbacks. Due to our incredible capacity for foresight, instead of just being anxious about those threats that exist in the present, humans can experience anxiety for all those possible threats which *might* jeopardize us in the future. For example, a human will not just experience anxiety over how he will procure his next meal, but how he will secure tomorrow's meals as well. And it's not just tomorrow's meals with which he's concerned, but all those that he will ultimately need to sustain himself far into the future, if not for the rest of his life. For this reason, though our capacity for foresight may serve as a great advantage to our species, it, at the same time, engenders a tremendous burden of anxiety.

In many ways, the anxiety function represents our primary defense in the continued struggle for survival. It is this anxiety function that keeps us ever-vigilant and alert, always on guard against the potential threats of hunger, dehydration, excessive heat or cold, strangers, disease, predatory animals, poison, sharp objects, fires, floods, droughts, hurricanes, the dark, etc., all things we have the unique ability to secure ourselves against, long before they represent an actual threat. It is this anxiety function that has motivated us to manufacture fire and electric light, to develop all sorts of medical technologies, to build dams, and structural fortifications, to erect silos, and to devise methods of refrigeration. Due to our enhanced capacity for foresight combined with the anxiety induced by our fear of future threats, we become obsessed with our futures. It is necessary that we be this way, for the minute we become lazy or lower our guards, we become vulnerable to a world of potential hazards and predators. The moment we become less anxious, we consequently become that much more endangered.

Whereas other animals may have claws or sharp teeth with which to protect themselves, humans possess the capacity for foresight. With our advanced capacity to envision our possible futures, humankind is that much more equipped to fortify itself against more threats than any other creature on Earth. But let me restate: "All that glitters is not gold."

– WHEN AWARENESS OF DEATH MEETS
THE ANXIETY FUNCTION –

"Anxiety is the state in which a being is aware of its possible non-being...The anxiety of death is the most basic, most universal, and inescapable."[21]   - Paul Tillich

"No one is free from the fear of death...The fear of death is always present in our mental functioning."[22]
-G. Zilboorg

"He that cuts off twenty years of life cuts off so many years of fearing death."   – Shakespeare

What becomes of this anxiety function when we combine it with our species' unique awareness of death? How are we to effectively utilize our capacity for foresight, when it is this same capacity that informs us of the fact that we are ultimately going to die? It is our capacity for foresight, combined with our capacity to experience anxiety, that keeps us perpetually vigilant and, consequently, alive. At the same time, however, we know that no matter what we do to fortify ourselves, because we are destined to die, our actions are all in vain. No matter how hard we work to provide ourselves with food and shelter, no matter what we do to protect ourselves, no matter how much we plan and prepare for our futures, we know that death is inevitable and inescapable. It is this awareness that strips the anxiety function of all its efficacy, in turn, stripping human beings of their capacity to effectively survive.

No other creature on this planet can comprehend the concept of its own existence. Consequently, no other creature can conceive of its own mortality. This seems to coincide with the fact that no other creature can comprehend the concept of its own future. Before us, all creatures lived in and for the moment. If an animal got hungry, it sought food. If it got tired, it slept. It lived and it died without one conscious thought regarding its own existence or non-existence. It had no conceptual awareness of its own possible future. The question, "What might happen to me *tomorrow*?" had never before been asked until man declared that such a day even existed.

"This time consciousness, which is possessed by no other species with such insistent clarity, enables man to draw upon past experience in the present and to plan for future contingencies. This faculty, however, has another effect: it causes man to be aware that he is subject to a process that brings change, aging, decay and ultimately death to all living things. Man, thus, knows what no other animal apparently knows about itself, namely that he is mortal. He can project himself mentally into the future and anticipate his own decease. Man's burial customs grimly attest to his preoccupation with death from the very dawn of human culture in the Paleolithic Period. Significantly, the burial of the dead is practiced by no other species. The menace of death is thus inextricably bound up with man's consciousness of time."[23]

To add insult to injury, not only are we aware that we must die, but we also know that death can come at any given moment. Regarding our futures, nothing is certain. We live our lives standing under the mythical sword of Damocles, waiting for that single hair, which holds inevitable death suspended directly above our heads, to snap at any moment.

Imagine how apparent this must have been for our earliest ancestors. How much security did primitive humans have that each day would not be their last? Imagine a time when there was hardly any medicine, when what may have seemed like a bellyache or toothache one day brought death the next. What constant dread and uncertainty must have plagued these people's existences. Among such nomadic dwellers, even the seemingly simple task of procuring one's next meal represented a potentially mortal chore. Whereas today one can merely pull up to the nearest drive-through to obtain one's daily ration of meat, these men had to go out with their crude hunting utensils and bludgeon some ferocious beast to death in order to procure their next meal. In such times, the threat of death was constant. Even so, with all of our modern conveniences and medical technologies, very little has really changed. Even with all of our advancements, there is still no escaping the fact that we are all destined to die, as well as the fact that death can come at any moment. Sure, we may live another twenty or thirty years longer than our predecessors, but what difference does that really make when measured against all eternity?

Living with the certain knowledge of death consequently leaves us in a constant state of mortal fear. At every given moment, we stand face to face with a mountain lion, staring straight into the jaws of death.

What this means is that, in light of our capacity for foresight, our awareness of inevitable death condemns us to live out our existences in a perpetual state of mortal fear and anxiety. The chief difference between our condition and that of the rabbit, as it stands face to face with the mountain lion, is that, unlike our own condition, the object of the rabbit's fear is tangible. Whereas the rabbit can escape the object of its fear, human beings cannot. Because we are aware of our own mortalities, our species is left incapacitated, in a state of constant mortal fear of an enemy that we can neither see, flee, nor defeat. In a sense, we are no better off than if someone had strapped a time bomb to us and then set it on a random timer that can go off at any moment within the next fifty or sixty years. What would we do in such a case other than to spend the rest of our existences in a state of perpetual mortal terror and dread, waiting for that ticking timebomb to finally detonate? How, I ask, is the human condition any different from this? The threat of death lurks around every corner, in every shadow, every meal, and every stranger. We don't know from where it will come, but that it inevitably will.

Just as there is no escape from death, there is no escape from the consequent anxiety that our awareness of death imposes upon us. With the awareness of our own mortalities, humankind had been left in a state of perpetual angst, dread, and despair, that which Kierkegaard called "The Sickness Unto Death." With the advent of self-conscious awareness, the anxiety function had been rendered impotent and with it our species.

It is the breakdown of this anxiety function in us that makes human beings the dysfunctional animals we are. In our frivolous attempts to either oppose or escape unavoidable death, we channel our energies into a morbid array of self-destructive behaviors. In our futile efforts to oppose the unopposable, we become that animal who will recklessly kill one another as well as our own selves. Unlike any other creature on Earth, we are capable of acts of suicide, sadism, masochism, genocide, torture, self-mutilation, drug abuse, along with a multitude of other disturbed responses, all of which I suggest result from our species' unique awareness of death. Because of our advanced capacity to comprehend the concept of death, humankind became the sick or, as Freud phrased it, the "neurotic," animal. \

Furthermore, in light of our awareness of inevitable death, life takes on a newfound sense of existential meaninglessness. Our struggles to survive become an exercise in futility. Between death's inevitability and all of the suffering we are forced to endure, we are compelled to ask: Why go on living? What's the point? How was our species to justify its continued existence in light of such a hopeless circumstance? Why struggle today when tomorrow we won't even be here? Under such preconditions, the motivating principle of self-preservation that had sustained life for all these billions of years no longer applied to our species. This was a whole new set of rules our animal was now playing by, and unless something could be done to ameliorate our species' pained, desperate, and pathetic condition, it might not have been long before our newly evolved animal would have succumbed to the forces of extinction.

– THE ADVENT OF THE SPIRITUAL FUNCTION –

"In order to counter this fundamental angst, humans are 'wired' for God."[24]   - Herbert Benson

So there we were, a newly emerging species with an unchallenged intelligence, one which had made us the most powerful creature on earth. Then, when everything seemed to be working out just fine, the inevitable took place - Man's intelligence backfired on him. For the first time in the history of life, an organic form turned its powers of perception back upon itself and became aware of its own existence. With the dawn of self-awareness, a cognitive revolution had taken place. With our newly acquired awareness of our own existence, we became equally aware of the possibility of non-existence. With this one cognition, the most powerful creature on earth was struck down, incapacitated by the crippling awareness of its own inevitable death.

Imagine how these first protohumans must have felt, suddenly cognizant of their own inevitable demise - naked, vulnerable, alone, defenseless against the threat of impending death, exposed before the void, unprotected by any "higher" force or being. If nature didn't provide this newly emergent creature with some mechanism by which to counter the anxiety - the horror - it's possible our species might have been slotted for extinction. Nature was going have to provide our animal with some sort of cognitive cushion if humans were to survive their unique awareness of death. Rather than being stricken by some devastating parasite or climactic change, Humankind had been assailed by a new environmental pressure, one which happened to originate from within our own brains. (After all, don't our own bodies constitute our physical environments?) Because of this new internal physiological pressure which came as a result of self-conscious awareness, it was now necessary that some advantageous adaptation be selected into our species if the hominid line was to survive.

In order for us to conform to this new environmental (physiological) pressure, the forces of natural selection could have affected our evolution in one of two ways. Essentially, our intelligence, which served as our greatest strength, was now jeopardizing our very existences. One evolutionary strategy that nature could have employed would have been to weed out the more intelligent members of our species, thereby leaving only the less self-aware members to survive. In other words, the forces of natural selection could have simply pushed us back a few stages in our evolutionary development and returned us to our former, less intelligent and, consequently, less self-aware states. The problem with this solution, however, was that hominids had already been stripped of most physical defenses. Without intelligence, human beings constitute one of the weakest and most vulnerable creatures on earth. Without our ability to make fire, for instance, mere insects, such as locusts, might have been able to outcompete us for our food supplies. So there we were, "the naked ape" - hiding out under rocks or in caves; without claws, fangs, wings, or venomous sting or discharge - nothing, save our vast intelligence. Had the forces of nature taken this from us, how would we have protected ourselves from the other species? Without our intelligence, Man is like a walking meal just waiting to be eaten.

Apparently, such a strategy probably wouldn't have worked in our favor. Instead, some mechanism needed to be installed in us that could allow our species to survive our mortal awarenesses without sacrificing our intelligence. But what kind of device could have accomplished this? What adaptation could possibly emerge in us that would relieve us of our inherent and incapacitating awareness of death without compromising our intellectual faculties?

Perhaps, at first, only those individuals whose cerebral constitutions somehow withstood the mortal anxiety that came with self-conscious awareness managed to survive. Nevertheless, something more was needed if the species, as a whole, was going to persevere.

Perhaps humankind's newly emergent awareness of death created so much tension in our animal that it induced a selective pressure on our cerebral physiologies, our brains. Just as environmental pressures transform entire species, why shouldn't these same pressures have been able to transform our organ, the brain? It was these same pressures that turned fish to amphibians, monkeys to men. Every single part of us from our opposable thumbs to our stereoscopic vision, from our language-enabling larynx to every single part of our brain evolved as the result of some environmental pressure. Shouldn't those same Darwinian principles that apply to all organic matter also apply to our cerebral evolutions as well?

Just as the industrial revolution induced a change in the Peppered Moth's environment, our species' awareness of death produced a change in ours. Whereas in the case of the Peppered Moth, the change in its environment was external, the change in ours was internal, taking place within our brain. Similarly, just as the change in the Peppered Moth's environment created an evolutionary tension that forced a change upon that creature's physiology, the change that had taken place within our own internal environments, within our brains, created a tension that forced a change upon ours. Our animal suddenly needed to be modified to meet the new demands placed upon it by its internal environment. What this meant was that those whose brains possessed some variation that incurred less psychic tension (anxiety), those whose lives were least disrupted by their awareness of death, were more likely to survive. Those more likely to survive, consequently, were more likely to pass whatever unique cerebral characteristic they might possess onto their offspring. As generations of these protohumans passed, those whose cerebral constitutions most effectively dealt with the anxiety that came as a result of their awareness of death were selected to survive. This process must have continued until a cognitive function emerged that altered the way these protohumans perceived reality by adding a "spiritual" component to their perspectives. Just as brains had evolved a musical, linguistic, and mathematical component, they developed a spiritual one as well.

In summary, I'm suggesting is that our species' awareness of inevitable death placed such a strong pressure on our cognitive evolution that over the course of millions of years - during the emergence of the hominids - nature selected those members who had developed a spiritual consciousness, a built-in perception that there exists an alternate, transcendental reality that supersedes the limitations of the finite physical realm, one that can only offer us pain, anxiety, and inevitable death.

Of those factors that may have influenced the evolution of a spiritual cognitive function, one I believe to have played a key role incorporates Man's unique capacity to enumerate.

Most animals possess an inherent comprehension of the dimensions of time and space. Because we live through time and in space, it is necessary that we possess such an awareness in order to survive. For instance, most animals possess an internal biological clock, one that serves to regulate an organism's behavior in relation to time. This biological clock will determine what time of the day or year an animal will forage, sleep, or mate, as a few examples.

Many animals rely, to a large extent, on their sense of sight for survival. Because our planet's lighting conditions are determined by the earth's rotation around the sun, this cycle plays a critical role in most animal behavior. Furthermore, because our planet's revolution around the Sun plays a critical role in the Earth's climate, this, too, will have a dramatic effect on much organic behavior. Because our environmental conditions are framed by time, it's necessary that most organisms possess an internalized biological clock that can help them effectively survive the Earth's cycles of light and climate.

Besides possessing an inherent perception of temporal events, all life forms possess a built-in mechanism that allows them to perceive the world spatially. Even a plant, though it may be rooted to the ground, engages in heliotropic behavior, the capacity to turn its leaves to face the sun. Because we exist within a three-dimensional (spatial) environment, most animals possess some combination of organs through which they can discern the directions of up and down, backward and forward, near and far. As mobile creatures, it would be impossible for any animal to survive without such spatial sensibilities.

Though most animals may possess a certain degree of temporal and spatial awareness, our species' capacity to comprehend both of these dimensions is by far the most advanced. Only humans can discern increments of time and space with such precision. By being able to apportion our world into such distinct spatial and temporal units, humans have evolved the capacity to enumerate objects, that is, to count*. Because we possess this particular "mathematical" cognitive capacity, humans are able to measure moments in time as well as units in space. Because our species possesses an enumerating or mathematical function, we alone have been able to navigate our way across the seas, continents, and, most recently, through extraterrestrial space. This capacity has also enabled us to construct immense architectural fortifications, countless machines and technologies,

---

*It was recently discovered that Rhesus monkeys possess the capacity to count groups of objects in consecutive order from one to nine. Here lies, yet another example of our evolutionary ancestors possessing an incipient talent for a predominantly human capacity.

along with formidable instruments of healing as well as destruction, all things which have, for better or worse, served to make us the most powerful and versatile creature on Earth.

Although such capacities may generally be to our advantage, just as our intelligence backfired on us, the advanced capacity to enumerate also affected us in a similarly hazardous way. I say this because inherent in our capacity to enumerate exists an awareness that this process has no finite end. No matter how big a number might be, we can always add one to it. Intrinsic in our capacity to add one and one exists the consequent capacity to conceptualize infinity. As only our species possesses this sophisticated a capacity to enumerate, only we can comprehend the concept of infinity.

In the same way that we can enumerate units in space, we can do the same with moments in time. Just as we can comprehend the idea that one plus one equals two, we can equally conceptualize the notion that this present day plus one more equals tomorrow. It is from this same capacity that humans may have gained their capacity for foresight, one that has enabled every human culture to devise some form of a calendar by which it measures time in days, seasons, and years.

Just as our enumerating capacity has enabled us to recognize that spatial dimensions possess no finite end, we can apply the same notion to temporal dimensions as well. Analagous to the way we can conceptualize infinity, we can do the same with eternity. Just as we can keep adding one unit to any spatial dimension, we can equally add one moment to any temporal dimension (this moment plus the next moment equals the moment after, etc,.).

With a capacity to conceive the eternal, not only can we conceptualize our own futures all the way to our inevitable deaths, but way beyond that, into eternity. Because we can comprehend the concept of eternity, our species must contend with an awareness that though we, our physical bodies, are temporal in nature, time itself will never end. With a conscious awareness of eternity, humans were suddenly forced to endure the notion of how infinitely brief life is. Whereas all the other creatures lived in and for the moment, we now had to measure our existence against the overwhelming backdrop of all eternity. Suddenly, humankind had to contend with an inherent sense of its own ultimate insignificance. In the words of the French philosopher and mathematician, Blaise Pascal, "The finite is annihilated by the infinite." Consequently, due to our capacity to grasp the eternal and the infinite, our animal now had to suffer a new anxiety, one which may have rivaled that which came as a result of our debilitating awareness of death.

Due to our capacity to comprehend the infinite and eternal, our species' newly emerged mathematical consciousness may have played as significant a role in the evolution of a spiritual function as did our awareness of death*. Not only did we now need to be protected from actual death itself but from all of the possibilities that might exist long after death. Suddenly, Man was aware that he might exist (or for that matter, not exist) for all eternity. But how? In what condition? Would eternity be a pleasurable or a painful experience? Would we retain our conscious identities and, if so, in what state? Would eternal existence be as replete with experience as this life or would it represent a state of absolute nothingness, of eternal non-existence? Furthermore, what might that even mean? As it is natural for humans to be concerned with the future, our species was suddenly condemned to spend our lives in search of knowledge of what might happen to us, not just during our lifetimes but long after death. With this new awareness, humans would now have to spend the duration of their existences living, no longer just in fear of death, but in fear of what may come after death, in fear of eternal non-existence.

Rather than allowing these fears to overwhelm and destroy us, perhaps nature selected those who processed their concept of time after death in an altered manner. Perhaps after many generations of evolution, humans were "rewired" to perceive infinity and eternity as an integral part of self-identity. Perhaps a neurological connection emerged in our species that compelled us to perceive ourselves as able to transcend the limitations of this finite realm and therefore able to transcend the limitations of physical death. This, I believe, repre -

*Mathematical or numerical consciousness apparently interacts with our spiritual consciousness. This relationship is made evident by the fact that every world culture has attributed spiritual significance to various numbers. Whether it be the jewish Kabbalists, the Pythagorean Greeks, the medieval alchemists, the Christian use of a holy trinity, the use of numbers in Aztec mythology, numerical references made in the *I, Ching*, or the general use of numbers employed by the variety of astrological and numerological belief systems, every world culture has maintained a belief that numbers can possess spiritual content.

sents the evolutionary force underlying the origin of our species' belief in an immortal [eternal] soul.

Once we perceived ourselves as possessing an element of the infinite and eternal within us, as clear as it was that our physical bodies would one day perish, we were now able to believe that our conscious selves, what we came to refer to as our souls, were immortal and would endure for all eternity. Herein lies the origin of our cross-cultural belief in immortality, in our inherent perception that our conscious selves transcend physical death. Once we came to perceive consciousness as eternal, we perceived death as just another life-passage in our eternal existence. Suddenly our animal was compelled to bury its dead with a rite that anticipated sending the deceased's immortal conscious self or "soul" onward to a realm of eternal existence. With this predisposition toward believing in our own immortality, our species was finally relieved of a large part of the burden that came with our fear of death. Humankind was saved.

But even if we were to live forever, what did that mean? Humankind still needed relief from the fear of the unknown. Would the afterlife be a place of eternal happiness or would it perhaps be even more painful and precarious than this stay here on Earth? Without our parents to protect us in the afterworld, humankind now needed eternal guidance and protection.

According to Freud, "God is the exalted father, and the longing for the father is the root of all religion."[25] Aware that death was not only inevitable, but that it could strike at any moment, human beings were reduced to a state of infantile helplessness, as vulnerable as the day they were born. And where do infants innately turn for protection? To their parents. However, not even one's parents can save one from death. As we become adults, we come to recognize that even our once-seemingly-omnipotent parents are defenseless against death. With this knowledge, where was humankind to find eternal guidance and protection? Desperately longing for eternal comfort and security, to whom or what was primal man to turn?

Our need for eternal protection may have caused nature to select an inherent awareness of a transcendental guardian into us, what eventually produced a series of neural connections which compelled the human animal to believe in a god or gods. As infants in the crib, when we experience pain or fear, we instinctively reach out to our parents for comfort and protection. It seems likely that our cross-cultural belief in a God represents an extension of that same instinct. As Freud expressed this same notion,

> "The derivation of religious needs from the infant's help-lessness and the longing for the father aroused by it seems to me incontrovertible, especially since the feeling is not simply prolonged from childhood days, but is permanent-ly sustained by fear of the superior power of fate. I cannot think of any need in childhood as strong as the need for a father's protection."[26]

Due to the selective pressure placed on our species by our aware-ness of eternity and death, humans had evolved an inherent belief in an all-powerful figure, an imaginary being whose infinite powers could protect us from death and all that came thereafter. Again, neu-rological connections may have emerged that forged a belief in spir-its and supernatural beings, beliefs that eventually came to manifest themselves in social institutions we refer to as religions.

> "If the brain evolved by natural selection...religious beliefs must have arisen by the same mechanism."[27]

In summary, what I am suggesting is that at some point in the last five million years or so, since the emergence of the hominids, an adaptation evolved that has enabled us to cope with our awarenesses of death, while, at the same time, allowing us to maintain our sense of self-conscious awareness. By having this cognitive mechanism instilled in us, we were now "wired" to perceive death as less deci-sive and therefore less threatening than it may have otherwise seemed. Once these cognitive phantoms that nature had instilled into our species' neurophysiological make-ups were given credence in the form of a socially accepted doctrine, we call religion, humans were better equipped to handle their fear of death.

> "One of the major functions of religious belief is to reduce a person's fear of death."[28]

> "Religion is a natural defense against man's knowledge that he must die."[29]

In essence, with the emergence of a spiritual function, our species could now survive. Sheltered from the perpetual threat of inevitable death, humans could proceed with the daily routine of maintaining their more "Earthly" needs. With the emergence of spiritual consciousness, our cognitive functioning had been stabilized to the extent that we could now go on living in a state of relative calm, even amid our awareness of inevitable death. This, I contend, is the purpose of the spiritual function. This is its rationale, its reason for being.

Again, if all this is true, it would suggest that God isn't a transcendental force or entity that exists "out there," but rather represents the manifestation of human cognition, a phantom generated from within the human brain.

In the remaining chapters, I will provide a variety of arguments meant to further support such a hypothesis.

THREE

THE SPIRITUAL EXPERIENCE

"Still there are moments when one feels free from one's own identification with human limitations and inadequacies. At such moments, one imagines that one stands on some small spot of a small planet, gazing in amazement at the cold yet profoundly moving beauty of the eternal, the unfathomable: life and death flow into one, and there is neither evolution nor destiny; only being."   — Einstein

"The mystical experience of God has certain characteristics common to all faiths."[30]   - Karen Armstrong

Having catalogued Man's universal spiritual beliefs and practices, there were still several other components to spiritual consciousness that I felt needed to be investigated. One such component came neither in the form of a belief nor a practice but rather as a sensation that appears to be cross-culturally experienced by our species. Moreover, it seems that there are certain acts which have the universal effect of evoking these sensations, ones each culture tends to characterized as being either religious, mystical, spiritual, or transcendental in nature.

In his book, *Civilization and its Discontents*, Freud discussed a letter written to him by his Nobel Laureate friend, Romain Rolland. In this letter, Rolland described sensations he had experienced that he felt represented the true source of all religious sentiment. According to Rolland, these sensations "consisted in a peculiar feeling, which he finds he himself is never without, which he finds confirmed by many others, and which he may suppose is present in millions of people. It is a feeling which he would like to call a sensation of 'eternity', a feeling as of something limitless, unbounded - as it were, oceanic."[31]

Romain Rolland was right in presuming his experience was one he shared with millions of people. As a matter of fact, research suggests that every culture from the beginning of our species' recorded history has described experiencing sensations almost identical to those articulated by Mr. Rolland. Whether it be in the form of Hindu samadhi, Sufi fana, Zen satori, or the born-again experience of the

Pentecostal Christian, every world culture has characterized a sensation we refer to as a "spiritual" or mystical experience.

So how should we define these spiritual experiences? What feelings or expressions besides those mentioned by Romain Rolland have been used to describe them? Before we can examine the nature of this cross-cultural phenomenon, we must first be able to identify it.

As mentioned, it seems that practically all attempts to define the spiritual experience have yielded striking similarities. Such phrases as: "a feeling of an indissoluble bond, of being one with the external world as a whole,"[32] "a 'higher' experience,"[33] "pure conscious experience,"[34] "cosmic consciousness,"[35] "feelings of unity,"[36] "a greater awareness of a higher power or ultimate reality,"[37] "diminishment or loss of sense of self,"[38] "dissolution of the normal ego; a new kind of ego functioning,"[39] "an altered perception of space and time; ineffable; appreciation of the wholistic, unitive integrated nature of the universe and one's unity with it,"[40] and "God-consciousness,"[41] are all examples of expressions that have been used to describe what we refer to as either religious, spiritual, mystical, or transcendental experiences. Furthermore, such experiences are usually described as evoking feelings of "equanimity; rapture; sublime happiness,"[42] "bliss,"[43] "ecstasy,"[44] "intense positive affect,"[45] "peace and joy,"[46] "a state of assurance,"[47] and "elation."[48]

In regard to the above characterizations of the spiritual experience, it seems there are two distinct types of sentiments described. The first set depicts the spiritual experience as one which is transpersonal in nature. Sentiments that describe a loss of sense of self or dissolution of normal ego functioning define the spiritual experience as one in which a person feels detached or dissociated from his usual ego boundaries, almost as if one's sense of self has been suspended or suppressed. In such cases, our normal ego boundaries seem to disintegrate leaving one feeling boundless, limitless, cosmic, or oceanic.

Unlike this first set of terms used to describe the spiritual experience, a second set portrays sentiments that point to a more sedative and sensual experience. Such terms as rapture, bliss, ecstasy, and inner peace all reflect of a sentiment of physical euphoria.

As a testament to the nature of these experiences, some cultures have even created words specifically meant to describe these particular sensations. The people of India, for instance, have a word, *Saccidananda*, which appears quite frequently in the religious and philosophical writings of this culture.

"This composite sanskrit word consists of three separate roots: Sat meaning existence or being; Cit, awareness and intellect; and Ananda, bliss."[49]

The fact that so many cultures have described experiencing these particular sentiments and in such similar terms suggests that this represents yet another cross-cultural characteristic of our species, i.e., another genetically inherited trait.

"Confirmation of the genuineness of mystical experiences is to be found in the high degree of unanimity observable in the attempts to describe its nature."[50]

Just as all cultures have described experiencing the sentiment of sadness, all cultures have equally described experiencing these spiritual sentiments. Furthermore, just as the experience of sadness is described by all cultures in such similar terms, the same is true of our spiritual sentiments. That all cultures, no matter how isolated, have described sadness in such a similar way indicates that this sentiment is not learned but is innate, a part of our inherent human experience. By the same logic, this should hold true for our spiritual experiences. Moreover, if our capacity to experience "spiritual" sentiments truly does represent an inherent characteristic of our species, this would further imply that such experiences must be generated from some physiological site in us, a conviction upheld by others.

"The mystical experience can be explained in physiological terms."[51]

Offering evidence in support of this notion, Andrew Newberg and Eugene D'Aquili at the Nuclear Medicine division at the University of Pennsylvania used Spect (single positron emission computed tomography) scans to observe changes in the neural activity of Buddhist monks engaged in the act of meditation. These experiments showed that while the monks were engaged in meditation, there was a sharp decrease of neural activity within their frontal and parietal lobes, providing physical evidence that spiritual experiences can be directly linked to one's neurophysiology.

Offering more evidence in support of this notion, at UC San Diego's Center for Brain and Cognition Research, Dr. V.S.

Ramachandran has recently found that a significant number of people who suffer from temporal lobe epilepsy describe undergoing intense spiritual experiences during their seizures, experiences that can be linked to neural activity within their temporal lobes. Because of this, scientists have now dubbed this part of the brain the "god module." Moreover, these epileptics not only exhibited an unusual preoccupation with religious concerns in their normal everyday lives but also displayed a heightened emotional sensitivity to religious language and icons. Similar to the way that music can trigger an emotional response in people, indicating there might exist a musical part of the brain, the fact that religious language or icons can do the same suggests there may exist a "spiritual" part of the brain as well.

In support of Dr. Ramachandran's findings, Jeffrey Saver and John Rabin of the UCLA Neurologic Research Center found that a significant number of the world's spiritual prophets and figures had exhibited symptoms indicating they may have suffered from this same type of epilepsy. The list they composed included, among others, Joan of Arc, Mohammad, and the Christian apostle, Paul.

Providing yet more tangible data to support a physiological explanation of spiritual experiences, the Canadian psychologist Micheal Persinger found that while using a transcranial magnetic stimulator (a helmet that shoots a concentrated magnetic field at a specific region within the brain) to excite his own temporal lobes, he experienced what he described as his first feelings of being at union with God.

Additional support for a genetic explanation of religion is offered by twin studies that showed, "Studies of twins raised apart suggest that 50 per cent of the extent of our religious interests and attitudes are determined by our genes."[52]

Freud, himself, speculated that spiritual experiences might be physiological and, consequently, genetic in origin. In discussing his thoughts on Romain Rolland's letter, Freud remarked that, "The idea of men's receiving an intimation of their condition with the world around them through an immediate feeling sounds so strange that one is justified in attempting to discover a *genetic* explanation of such a feeling."[53]

- ORIGINS OF THE SPIRITUAL EXPERIENCE -

If we presume that this sensation we cross-culturally define as a spiritual, religious, mystical, or transcendental experience represents a genetically inherited trait, we must, as always, ask why? Why does our species have a tendency to experience this particular sensation? What is its purpose, its function? How might it serve to enhance our species' survivability? Again, if this experience provided no specific purpose or function, it's highly unlikely that such a trait would have emerged in us.

As discussed earlier, the spiritual function probably evolved in our species in response to the emergence of human self-conscious awareness which brought with it, as an unfortunate side-effect, an awareness of our own mortality. As a result of this new form of consciousness, the human animal would have to live in a state of constant mortal peril and dread unless some adaptation could somehow cushion the painful effects of this awareness. Again, if the anxiety evoked by our mortal awarenesses wasn't somehow alleviated, our species might not have survived.

Our spiritual function compels us to perceive the existence of an alternate transcendental reality. Some of the ways our spiritual consciousness manifests itself is through a belief in a supreme transcendental being, a soul, and the continuity of that soul in what we call an afterlife. As a result of these inherited cognitions, human beings maintain a belief that they are immortal. In perceiving ourselves as immortal, we are relieved of a great deal of the psychological strain that comes as a result of our unique awareness of death. But believing in the existence of a spiritual reality is one thing - to experience it, quite another.

Imagine a person who has been shipwrecked on a desert island, abandoned to the elements without a trace of food. Five days later, he lifts his aching head to notice a rescue ship heading toward him. Knowledge that food is on the way would certainly relieve this starving person of a great many of his anxieties. Nevertheless, believing that food is on the way and experiencing the food itself are still two different things.

Just as fear of starvation can evoke feelings of anxiety, so can the fear of death. Furthermore, in the case of our shipwrecked and starving survivor, though believing that food is on its way may alleviate some of his anxieties, it is the actual experience of eating that will diminish most of his actual physical discomfort or pain.

Analogously, though believing in a spiritual reality might help to relieve us of some of our mortal fears and consequent anxieties, I suggest that we were instilled with an added capacity to experience these euphoric "spiritual" sensations in order to complement our beliefs. Again, to perceive a spiritual reality is one thing, to experience it, quite another. This, I believe, is the function of the spiritual experience, to help alleviate some of the excess physical strain that comes as a result of our species' awareness of death.

My next question was: Physiologically speaking, what is a spiritual experience? What physical mechanism could be responsible for this particular set of sensations that seems to be exclusive to our species?

– THE EGO FUNCTION –

"The self is a relation which relates itself to its own self."[54]
- Soren Kierkegaard

There is not a healthily functioning human who cannot recognize his or her own reflection. Though most other animals can identify one of their own species, only humans can recognize themselves*. Only humans possess a developed sense of self-conscious awareness.

As the only species to possess the capacity for self-awareness, it must represent a trait that emerged sometime during the evolution of the hominid line, animals that emerged from the primates (which didn't possess a developed sense of self) and of which we are the last surviving species. Since self-awareness represents a cross-cultural characteristic, we can presume that it represents the cognitive manifestation of yet another genetically inherited trait. This would suggest that there exists a physiological site or group of sites within the brain, from which self-conscious awareness is generated. I will refer to this site, in that it provides us with a coherent sense of self, as the ego function.

As is true of any of our cognitive functions (including the spiritual), the ego function is comprised of a group of interactive cognitive parts or processors. Before I discuss the ego function's evolution-

---

*Though only humans possess a capacity for self-awareness, evidence indicates that apes also possess self-perceptual capabilities, but to a much more limited extent. In one experiment (Gallup, 1970), chimpanzees were housed in individual cages with a full-length mirror left outside facing them. For the first few days the chimps screamed at the sight of their own reflections, made threatening gestures, and generally behaved as though they were being confronted by another chimpanzee. Several days later, however, the chimps' behavior changed. Instead of acting as if their reflections were another animal, they began to use the mirror as a tool with which they groomed themselves, as humans do when we, for instance, comb our hair. In some cases, the chimps were seen using the mirror to pick food particles from their teeth. (Monkeys, on the other hand, even after hundreds of hours of being left in the same set-up, showed no signs of self-recognition.) Once again, given the chimps' evolutionary proximity to our own species, it would make perfect sense that they would show such incipient self-perceptual capacities.

ary significance or its relationship to a spiritual experience, I will first discuss the nature of some of these interactive parts.

Mapping out the physiological components of human consciousness has presented science with one of its most enduring and perplexing challenges. What is consciousness? What is self-conscious awareness? Where does it originate? How does it operate? Is there a neural basis for self-conscious awareness or is this just another function of our transcendental and eternal souls?

One of the chief components underlying self-awareness involves something called "episodic" or "autobiographical" memory. Episodic memories are those related to one's past experiences, and can be linked to the brain's hippocampus where it is believed such memories are processed. We believe this because damage to the hippocampus has been cited to precipitate a variety of amnesias causing individuals to forget everything from their names, careers, and addresses, to the fact that they are married or have children. According to cognitive scientist David Noelle,

> "Some amnesics can recall events from early in life, but fail to form new memories for life events. Thus, they may have a coherent sense of self but might feel as if no time has passed since their damage appeared. Other amnesics seem to retain no memory of their past at all. They emotionally report a sense that today is the first day of their lives...that they have just become conscious. Our memories apparently play an important role in constructing a sense of ourselves as unified entities persisting through time. Without these memories, our sense of self seems somewhat disrupted or disturbed."

As stated by V.S. Ramachandran in his book *Phantoms in the Brain*, "If you lost your hippocampus ten years ago, then you will not have any memory of events that occurred after that date."[55]

Another component to play a critical role in our sense of self-awareness takes place in what we call "body-image" or "self-representation." Body-image constitutes that part of the human conscious experience through which we are aware of our physical presence. This part of the human experience also involves what is called "body-consciousness." For instance, even if my eyes are closed, I am still aware of the fact that I have raised my right arm, if I have done so. It has been suggested by Dr. Ramachandran that this particular form of consciousness can be attributed to the workings of the right parietal lobe. This hypothesis is based on the fact that people with damage to their right parietal lobes develop an altered sense of body consciousness. For example, many people with right parietal lesions, who are paralyzed on the left side of their bodies often deny that they are paralyzed and will even describe imaginary movements such as that their left arm is waving about even though that arm is immobile. This tendency to imagine illusory or phantom body movements, or to *confabulate* as it is referred to by neuroscientists, is a common characteristic of patients with right parietal lesions. (If our brains can make us believe in phantom limbs, why not phantom beings?)

Cotard's syndrome represents yet another example of a cognitive dysfunction in which the victim suffers from an incapacity to comprehend his own being. As a result of damage to the brain's amygdala, a person may feel alienated or dissociated from his own body or body parts. Someone suffering this syndrome might, for instance, look down at his own arm and suggest that it doesn't feel as if it belongs to him. Such dysfunctions as these indicate that the components of self-conscious awareness are inextricably linked to one's neurophysiology.

Another component of self-conscious awareness exists in our capacity to make choices or decisions. This function, referred to as one's executive processor, has been "associated with the limbic system including parts of the anterior cingulate gyrus. This process connects perceptual qualia [subjective experience] with specific emotions or goals, enabling one to make choices."[56] Furthermore, "when the amygdala and anterior cingulate gyrus are disconnected, disorders of 'free will' occur."[57] People who suffer from such "free will" disorders may feel incapable of making decisions. Those who suffer this syndrome may come across as emotionally paralyzed, their movements often appearing forced like that of an automaton.

Based on such data, it would appear that human self-conscious awareness is linked to one's neurophysiological make-up as opposed to being representative of some incorruptible, immutable transcendental soul.

With such an organic interpretation of self-identity, the reader should realize that when I refer to an ego function, it is not to be confused with Freud's or Jung's definitions of the ego. Though I agree with Jung that the ego, or what he referred to as "ego-consciousness," represents that part of us from which our sense of self is generated, Jung viewed consciousness as a manifestation of the ethereal *mind*, whereas I see consciousness as a purely physical phenomenon, the manifestation of electro-chemical impulses being transmitted throughout the brain. One could therefore say I am seeking to biologize Jung's conception of ego-consciousness.

But before we take such an organ as an ego function for granted, we must first ask: If our capacity for self-awareness represents the manifestation of a physiological process, what is its purpose? How might such a function increase our species' survivability? Again, if no such purpose can be determined, it is not possible to justify the existence of such a trait.

As infants, we do not yet possess a developed sense of self. At this early stage in our development, a human being cannot distinguish its own existence from the world around it. As Freud expressed it, "An infant at the breast cannot distinguish his ego from the external world as the source of sensations flowing in upon him."[58] What this means is that when we are born, the ego function, like our language functions, exists in a latent stage of development. Therefore, instead of being fully developed at birth, our sense of self-awareness - our ego function - is apparently something that emerges during our post-fetal developments.

The developmental psychologist Jean Piaget also proposed that humans are born without any recognizable sense of self. Studying the cognitive development of children, Piaget observed that before the age of two, children possess little, if any, sense of self-awareness. Piaget classified this pre-self-aware phase of our existence as the "preoperational stage" of human development.

According to Piaget, it is between the ages of two and seven, during what he referred to as the "operational stage," that a child learns to recognize his own image, his own self. It is during this stage of development that a child recognizes himself as an autonomous being, separate and unique from his mother and the rest of the world.

As the child becomes conscious of his autonomy, he develops a sense of self-responsibility. He realizes that he must learn to fend for himself. It is during this stage that a child also learns to feed himself, wash himself, and go to the bathroom by himself. Perhaps, it is during this stage of human development that one's executive processor or anterior cingulate gyrus becomes activated. And so, slowly but surely, we grow from utterly dependent to independent (or, at least, interdependent) beings.

As the child's sense of self unfolds, he develops an instinct for self-preservation, the desire to sustain and to protect his newfound self. The stronger his sense of self, the more he is motivated to care for himself. As a result of our capacity to recognize ourselves, we have become the only species in which an individual can develop a genuine bond with himself. We are consequently nature's first narcissistic creatures, the first animals to possess a capacity for self-love. In a sense, one could say that we develop the equivalent of maternalistic feelings for ourselves. And so, with the same fervor and intensity with which a mother will love, care for, and defend her young, human beings can love, care for, and defend themselves. This, I believe, constitutes one of the chief advantages of self-awareness.

It is for this reason that the operational stage plays such a critical role in our emotional development. The conditions under which a child is raised during this stage (often referred to as one's formative years) will determine the manner in which he will learn to perceive himself. If a child is raised in a nurturing and loving environment, he will develop a positive self-image, in which case he will learn to love and cherish himself. The more a human loves and cherishes himself, the more effectively he will fend for himself.

If, on the other hand, a child is raised in an unhealthy environment, he will likely develop a negative self-image that may eventually foster a host of unhealthy and self-destructive tendencies. Such unhealthy tendencies we call neuroses. Neuroses are therefore the behavioral consequences of an unhealthily developed ego function.

Yet another benefit of self-awareness is that it grants us the ability to modify ourselves. Because we can perceive ourselves, we can recognize our own shortcomings. This affords us the capacity to turn our weaknesses into strengths. For example, though humans aren't born with the capacity to fly, should we perceive this as a shortcoming, we can build ourselves flying machines. Though we might not be born the fastest creatures on earth, by recognizing this as a shortcoming, we can build ourselves racing machines. Should another ice age strike, we won't need to wait millions of years for nature to select thicker coats of hair for us but can, instead, sew ourselves new coats within a few hours' time. On a more individual level, should a person recognize, for example, that he is dangerously overweight, he can diet. In this way, our species, and ours alone, has the capacity to modify itself, to compensate for any physical deficit and, consequently, to transform any shortcoming we might possess into a potential strength, thus rendering us the most versatile and resilient of all Earth's creatures.

So how does the ego function operate? The ego function acts as the body's control center or, as referred to earlier, its executive processor. If the body were a ship, the ego would be its captain. If the body is our temple, the ego functions as high priest. Whereas the heart is responsible for the pumping of blood, the ego is responsible for the supervision of our body's entire upkeep. It does this by acting as our body's personal manager, that part of us which is responsible for making all decisions. Should I seek food first or shelter? Shall I turn right or left at the next corner? All such choices are made not by our kidneys, livers, or language centers within the brain, but by that part or parts from which our sense of self is generated.

As I've stated, the ego function is responsible for the body's entire upkeep. For example, when we feel hunger, it is my ego that tells me, "*I* must provide food for *myself.*" As the manager of our physical beings, it is, consequently, the ego that must bear the brunt of all of our needs and responsibilities. When hunger must be assuaged, it is not the heart's, or the stomach's, or the kidney's responsibility, but the ego's to find the body its next meal.

When an individual feels pain, it is his or her ego that suffers. For example, if I were to get stuck in the hand with a pin, it is not my hand that suffers but "me," my ego, that bears the pain. Nullify a person's ego mechanism and you can turn him into a human pin cushion, and he won't feel a thing. My hand doesn't experience pain; *I* do! It is not my tongue that tastes the apple but *me,* my ego that does.

It is therefore the ego that is accountable for everything we do. Should I need to procure a meal or find shelter, it is *I*, my ego, that bears the brunt of this responsibility. It is therefore *I,* my ego, that must bear the brunt of all my anxieties. This includes that most debilitating anxiety of all which comes as a result of our unique awareness of death.

As is the case with any organ, when physical strain becomes too great, that organ is susceptible to suffering a mechanical breakdown. If I lift too much weight, I may tear a ligament. If I overexert my heart, I may suffer a heart attack. For every physical body part we possess, there exists a threshold of strain before that part will break. Consequently, if the ego is physiological in origin, it means that, just as with any other internal mechanism that is overloaded with stress, our ego mechanism can and will break. Consequently, if our ego mechanism didn't possess some means by which to relieve itself of the excess strain [anxiety] that comes with our awareness of death, it would be at risk of suffering a physiological breakdown. And when the ego breaks, all is lost. After all, what good is a ship once its captain has been lost?

What happens when our egos must bear the overwhelming strain that comes as a result of our species' awareness of death? Imagine having to experience one's entire life under the same conditions that a rabbit experiences when cornered by a mountain lion - its body pumped with adrenaline, its heart palpitating, its muscles tensed, its brain surging with painful fear and anxiety. Imagine having to experience this same anxious strain all day, every day, for the rest of one's life. Under such stressful conditions, how could one survive? How would one be able to perform any of life's normal daily functions? It would be impossible. Under such stressful conditions, one can only hope to either fight or to escape the object of one's fears. But in respect to the threat of impending death, there is no escape. Instead, we are left in a perpetual state of existential paralysis, unable either to fight or to escape the object of our fear.

Imagine the burden such conditions must have placed on our newly emerged ego mechanisms - exactly the type of undue strain that would render any physiological function susceptible to collapse. If our egos were to continue to function under such conditions, some device had to be installed in us that would relieve us of at least some of this undue strain. Had nature not provided us with such a mechanism, it's possible that our species would have suffered the kind of physiological breakdown that would have rendered us extinct.

It was at this point in our species' evolution, at the dawn of self-conscious awareness, that the forces of natural selection provided us with a mechanism by which our ego functions could endure the overwhelming strain that came as a result of our debilitating awareness of death. I refer to this mechanism as the "transcendental" function.

– THE TRANSCENDENTAL FUNCTION –

"The peculiar structure of the human ego results from
its incapacity to accept reality, specifically the supreme
reality of death."[59]   - Norman O. Brown

"Sometimes as I drift idly on Walden Pond, I cease to live
and begin to be."   - Henry David Thoreau

In order to save our ego functions from the severe strain caused
by our constant awareness of death, nature could have done one of
several things. It could have displaced the strain onto some other part
or organ, something that would only have proved to be equally dam-
aging (This tends to happen to a certain extent anyway, as psychological stress has been cited to play a key role in the origin of a number of physical illnesses). As I mentioned earlier, nature could have
also weeded out the more intelligent of our species, therefore, eradi-
cating our capacity for self-conscious awareness and with it our
awareness of death, but this would have proven ineffective as well.

Another strategy nature could have employed would have been
to select a mechanism that could suspend or suppress our ego functions long enough to relieve us from some of the strain of our excess
anxieties. If nature could provide a mechanism that would temporarily suppress our ego functions, it might assuage the burden of anxiety
suffered by that organ enough to allow it to continue to function.

If we recall the expressions used to describe spiritual experiences, there was an entire selection that involved the suppression of
one's ego function. Such terms as "loss of sense of self" or "dissolution of one's normal ego boundaries" reflect states in which the ego
function is temporarily suspended. With the ego temporarily shut
down, there is no longer a coherent *I* through which to experience
pain or anxiety. Instead we are left feeling egoless, detached from any
sense of self, a state universally depicted as boundless, limitless, lib-
erating, oceanic.

With our egos temporarily suppressed, we retreat to a state similar to the one in which we were born, one in which we couldn't differentiate between our own internal reality and the world around us.
As Freud expressed it:

"Our present ego feeling is, therefore, only a shrunken residue of a much more inclusive - indeed, an all-embracing - feeling which corresponded to a more intimate bond between the ego and the world around it. If we may assume that there are many people in whose mental life this primary ego-feeling has persisted to a greater or lesser degree, it would exist in them side by side with the narrower and more sharply demarcated ego-feeling of maturity, like a kind of counterpart to it. In that case, the ideational contents appropriate to it would be precisely those of limitlessness and of a bond with the universe - the same ideas with which my friend elucidated the "oceanic" feeling."[60]

With our ego functions suppressed, we experience sensations of "being one with the external world as a whole," of "pure conscious experience," of "cosmic" or "God" consciousness. And so, with our captain temporarily relieved of its duties, all of our anxieties are temporarily held in abeyance. In such a state, we are free from all sense of personal responsibility, egoless, immune to all physical pain and suffering. This, I imagine, is why the spiritual experience is often described as generating feelings of "euphoria," "rapture," "bliss," or "tranquillity."

And how are these spiritual experiences evoked in us? It seems that through acts of meditation, prayer, Yoga, dance, and even chant, we possess the capacity to trigger these "spiritual" sensations. To refer to a letter written by another of Freud's colleagues:

"Through the practices of Yoga, by withdrawing from the world, by fixing the attention on bodily functions and by peculiar methods of breathing, one can in fact evoke new sensations and coenanasthesias in oneself, which he regards as regressions to primordial states of the mind which have long ago been overlaid. He sees in them a physiological basis, as it were, of much of the wisdom of mysticism."[61]

In a normal wakened state, humans experience a brain wave frequency of about 13 cycles per second - which is referred to as a Beta wave. When we close our eyes and focus our attentions inward, that is, when we meditate, our brains shift to an Alpha state of 8 to 12 cycles per second. Apparently, just as the scientists at the University of Pennsylvania had proven through the use of SPECT scans, the act of meditation affects us physiologically. In addition, it has been shown that when a person is in the midst of an Alpha brain wave state, there is a tendency to be less responsive to physical pain.

We are all familiar with the notion of people claiming to be impervious to normally painful stimuli while undergoing such meditative or trance-like states. Whether we've seen this demonstrated in the form of someone lying on a bed of nails or walking across hot coals, the evocation of meditative practices seems to make us at least partially immune to physical pain.

According to studies done on Yogis, those practicing meditation "claim to reach a state [known as *mahanand* (ecstasy)] that surpasses the experience of pain."[62]

And how is it possible that we can immunize ourselves from physical pain? It's because when we suppress our ego functions there is no conscious self through which to experience pain or anxiety.

Another common symptom of a spiritual experience involves feelings of timelessness and spacelessness. In order to make sense of these sensations, I turned once again to Piaget's developmental theories. According to Piaget, a child cannot distinguish between proportions of time and space before reaching the age of seven, or what he referred to as the stage of "concrete operations." The fact that this capacity to distinguish proportions of time and space emerges in all humans at approximately the same age indicates that this ability represents a part of our species' natural cognitive development. This would imply that our species must possess some genetically inherited physiological site in the brain from which such temporal and spatial awarenesses are generated.

Because the stage of concrete operations occurs after the development of our ego functions (which, in Piaget's terms, takes place during the operational stage), perhaps our capacity to distinguish proportions of time and space is somehow inextricably linked to our capacity for self-awareness. Afterall, don't we perceive ourselves as existing within a certain temporal and spatial framework? Don't we view ourselves as existing either here or there, then or now? This would make sense in light of the fact that, by suppressing our ego functions, we seem to concurrently lose our capacity to comprehend dimensions of time and space. By suppressing our egos, we revert back to the preoperational stage of development (to again use Piaget's terms) when neither ego-consciousness, nor time and space consciousness existed in us. By evoking the transcendental function, our capacity to comprehend notions of self, time, and space are all diminished. This is why the spiritual experience is universally described as simultaneously evoking feelings of selflessness, timelessness, and spacelessness. With our sense of self, time, and space suppressed, we tend to interpret our experiences as ethereal, numinous, what we describe as being states of "cosmic" or "God" consciousness.

In essence, what I'm suggesting is that all such "spiritual" sentiments that make us feel as if we are "one with the universe" are really nothing more than the perceptual manifestations of a strictly neurophysiological process. By engaging in acts of dance, chant, prayer, and meditation, our species possesses the unique capacity to suppress the physical mechanics of our ego functions which, in turn, trigger sensations we interpret as being spiritual, mystical, religious, or transcendental in nature. Because every culture has described experiencing such feelings, and all in such similar terms, leads me to believe that what I call the transcendental function represents yet another evolutionary adaptation selected into our species.

With this, I hope to have demonstrated that spiritual experiences are not divine in nature, but rather represent the manner in which we interpret a specific series of electrical signals as they are registered in our brain. As succinctly expressed by the neurobiologist, Steven Rose,

> "It is highly probable that in due course it will be possible to explain the 'mystic experience' in terms of neurobiology; it is highly improbable that neurobiology will ever be explained in terms of 'the mystic experience.'[63]

FOUR

DRUG-INDUCED GOD

"Psychedelic drugs have been used to stimulate religious experience since the dawn of history."[64]   – C.D. Batson

"Religion is the opiate of the masses."   – Karl Marx

Besides various meditative techniques, many world cultures have described using psychedelic drugs as yet another means through which to induce a spiritual experience.

"The use of psychedelic sacraments in shamanic and religious practices is found throughout history. The word entheogen, used to describe certain plants and chemicals used for spiritual purposes, emphasizes this long established relationship."[65]

From the sacred drink of Soma used by the Vedic Hindus, the morning glory seeds and mescaline ingested by Native Americans, the sacred mints of the Greek mystery religions, the use of cannabis by the Sythians, and the Yaje of the Amazonian jungle peoples, to the Iboga of the peoples of equatorial Africa, all represent examples of drugs that have been used by various cultures to evoke a spiritual experience. Because of the universal nature of this phenomenon, the word "entheogens" (meaning "God generated from within") has been created to describe this class of "God" inducing drugs. To the ancient Aztecs, the connection between entheogens and the spiritual realm was so clear that they referred to peyote as the "divine messenger" and psilocybin as "God's flesh."

It is so widely recognized that certain drugs can stimulate a spiritual experience that some secular governments which normally forbid the use of drugs have allowed for the use of certain of these entheogens when ingested as a religious sacrament. "In 1994, the U.S. government enacted the American Indian Religious Freedom Act Amendments, providing consistent protection across all fifty states for the traditional ceremonial use of peyote by American Indians... In its report on the 1994 legislation, a U.S. House of Representative's committee reported that 'peyote is not injurious,' and that the spiritual and social support provided by the Native American Church [NAC] has been effective in combating the tragic effects of alcoholism among the Native American population."[66]

From William James' experiments with nitrous oxide to Aldous Huxley's experiments with Lysergic acid (LSD), it is a widely accepted fact by scientists from a range of disciplines that certain plants and/or chemicals can induce experiences indistinguishable from certain mystical states. But how is it that drugs can do this? How is it possible that chemicals can have the capacity to induce something as allegedy ethereal in nature as a spiritual experience? What does this say about such drugs? Or, more significantly, what does this say about the human spiritual experience?

In order to answer these questions, let's start by taking a look at the drugs themselves. As we know, all drugs, including the psychedelics or entheogens, as they are now called, are always the same in regard to their molecular structure. This is true of any drug. For example, on a molecular level, aspirin is always aspirin; penicillin is always penicillin. Accordingly, the same rule applies to each of the various entheogenic drugs as well. In other words, the chemical make-up of any entheogenic drug represents a constant. The molecular structure of a molecule of LSD is the same whether ingested in Bangkok or Bolivia, at sea level or on top of the Himalayas.

The same can be said, more or less, about human physiology. Granted, though there is a certain degree of variance among individuals within our species, underlying this diversity is a distinct physiological uniformity. Since we are dealing with two constants (same drug, same physiology), it is no surprise that entheogenic drugs should have a similar effect on most individuals, regardless of any cultural distinctions. This still leaves us with the crux of the problem which is, why do these drugs have this particular effect on us? Why do they have a distinct tendency to elicit what we refer to as spiritual/mystical/transcendental/religious experiences?

No drug can elicit a response to which we are not physiologically predisposed. Drugs can only enhance or suppress those capacities we already possess; they cannot, however, create new ones. For example, the fact that we possess the capacity for sight - that we possess the physical hardware to see - means that it is within the realm of possibility that a drug would be able to either enhance or suppress one's visual capacities. The fact, however, that we do not possess the physical hardware to telekinetically levitate objects means that no drug can ever enhance or suppress our non-existent powers of levitation. Again, a drug can only affect us as much as we possess some physiological mechanism that might be receptive to that drug's particular chemistry.

The fact, for instance, that novocaine has the universal effect of desensitizing one to pain means that we must possess pain receptors that are capable of being chemically suppressed. In the same way, the fact that psychedelic drugs have a cross-cultural tendency to stimulate experiences we define as being spiritual means we must possess some physiological mechanism whose function is to generate this particular type of experience. If we didn't possess such a mechanism, there's no way that these drugs, on their own, could stimulate these sensations in us. In short, the fact that there exist drugs that can induce a spiritual experience in us, again, supports the notion that our spiritual sensibilities must be physiological in nature.

FIVE

THE PRAYER FUNCTION

"People who reported increased spirituality described...the presence of an energy, a force or power – God – that was beyond themselves. It was the people who felt this presence who rated the greatest medical benefits...regardless of their faiths."[67]   – Herbert Benson

The notion that the act of prayer possesses distinct healing properties is nothing new. It has been written about and documented many times over. As a matter of fact, there are whole shelves in today's book stores devoted to this particular topic, usually under the heading of what are dubiously called the "New Age" sciences. I will therefore not reiterate such texts but will only address the issue insofar as it relates to a physiological interpretation of this same cross-cultural phenomenon.

From every corner of the globe a variety of cultures have spoken or written of the healing properties of prayer.

"Many cultures have named and believed in a mysterious healing energy. The ancient Egyptians called it 'Ka', the Hawaiians, 'Mana', the Indians 'Prana.'[68]

The fact that this phenomenon seems to be cross-cultural in nature suggests that we are dealing with yet another genetically inherited characteristic of our species. The fact that so many cultures have documented the healing properties of prayer also implies that our species must contain some internal series of mechanisms that are physiologically responsive to this particular act. Just as meditation seems to trigger a distinct set of physiological responses in us, the same seems to hold true for the act of prayer. Just as the act of meditation cross-culturally stimulates what we refer to as mystical or religious experiences, the act of prayer seems to cross-culturally elicit a healing response.

And when we pray, to whom is it we are usually praying? Whom else but to our gods? In other words, this prayer function is apparently directly related to our spiritual function.

When we pray, we are physiologically affected in such a way that we can be cured of certain physical debilities or illnesses. Not only this, but there is evidence that prayer can also expedite the time it takes to recover from illness or surgery. What this demonstrates is that somehow, through the act of prayer, humankind possesses the capacity to heal wounds, cure illnesses, and prevent disease. But how might such a mechanism work in us? How is it that the act of prayer possesses such unusual healing properties? Is this the work of miracles or is it simply another physiological response to a specific stimulus?

As we know, the human body is an interactive network of organs. If one organ is not doing its job properly, the rest of the body will suffer. It is the job of the kidneys, for instance, to filter toxic wastes from our body. If the kidneys are not functioning properly, the unfiltered toxins will have an adverse effect upon the rest of the body. As another example, when a person has a healthy heart and circulatory system, the rest of the body's tissues and organs are provided with an effectual supply of oxygen, allowing each of those parts to operate at maximum capacity. An inefficient heart or circulatory system will therefore have an adverse effect on every other part us.

As this principle applies to all of our organs, it must also apply to our brain. Therefore, if the brain, which is the body's control center, is not functioning at maximum capacity, neither will the rest of the body. Of the many functions that the brain provides, one is to channel our anxieties. If the brain is not performing this function properly, this, too, will have an adverse affect on the rest of the body.

In its healthiest form, anxiety works to one's advantage as it is meant to heighten one's capacity to respond to an urgent situation. In its unhealthiest form, however, anxiety has been attributed to be the cause of everything from panic attacks, nausea, sleeplessness, diarrhea, hair loss, disturbance of sleep patterns, ulcers, palpitations, trembling, headaches, loss of appetite, and even cancer, to name a few examples, all things which place yet even more strain on the body.

As I've already mentioned, it is the ego function's duty to oversee the upkeep of the rest of the body. It is, therefore, also the ego function that must carry the burden of all those subsequent anxieties that come with this vast responsibility. Consequently, there is a great deal of strain placed on this physiological part of us. For this reason, if the ego mechanism cannot effectively displace such strains, it will not be able to perform at maximum capacity.

In keeping with the principle that the body works as an interactive network of organs, if the ego mechanism is not functioning at maximum capacity, neither can the rest of the body. Any strain that the ego mechanism cannot properly handle may end up being displaced upon some other part of us.

As most of us possess some part of our body that is more vulnerable than the rest, it is often this part or organ that will suffer the effects of such strains. This same principle also applies to those organs that are vulnerable because they might either be ailing or in the process of recovering from some sickness or surgery. As these organs are either indisposed or in the process of healing, they are most vulnerable to the adverse effects of our excess anxieties.

As the body represents a nexus of interactive parts, it should come as no surprise that excess anxiety, which will weaken our ego function, would be able to bring about any number of illnesses. It should therefore make equal sense that by reducing our anxiety levels, we should be able to reduce our chances of incurring illness. Furthermore, if anxiety can weaken any part in us, it can also weaken our immune systems. Therefore, by reducing our anxiety levels, we should be able to optimize our immune systems, which, in turn, will expedite the healing or recovery process even more.

So how might prayer be related to our spiritual function? It seems as though humans have an inherent propensity to believe in supreme supernatural beings who have powers that far surpass our own. In times of adversity, humans have a tendency to turn to these "higher" powers for their aid or assistance. Because we believe that these same gods have created all that exists, including our own selves, we believe that, as our creators, they possess certain maternalistic/paternalistic feelings for us. For this reason, we believe that when we solicit our gods for assistance, that is, when we pray, those same gods will come to our assistance. Just as our parents were there to care for and to protect us as children, we instinctively believe that our gods are there to care for and to protect us as adults. Because we instinctively believe in the existence of spiritual forces that possess both the power and the inclination to assist us, we are compelled to pray to these forces. Because we believe our solicitations will be answered, this helps to diminish our anxiety levels, thus relieving us - our ego functions - of a certain amount of excess psychobiological strain. By diminishing our anxieties, that is, by relieving our ego functions of any excess strain, the rest of our bodies, including our immune systems, with all of their regenerative powers, can function at maximum capacity. With this accomplished, not only will there be less strain displaced on an ailing or recovering organ, but with our immune systems able to operate at a greater capacity, they, too, can more effectively help to implement the healing process. This, I believe, represents the underlying reason that the act of prayer engenders the healing properties it does.

As the body's control center, the brain plays an influential role in nearly every bodily function. Since a strained ego mechanism can interfere with the workings of any other part of our bodies, it can also adversely affect all of our other brain functions.

When we walk, for instance, it is not because our legs have decided to move of their own accord but because we, by virtue of our brains, have directed them to do so. A distressed brain could therefore theoretically interfere with the operation of one's entire nervous system, thus, for example, rendering a person with perfectly functional legs paralyzed. This is the underlying cause of what are referred to as psychosomatic illnesses. These illnesses originate, not in the afflicted body part itself but from within the workings of the brain. It is for this reason that people who suffer from psychosomatic illnesses are usually cured, not with the aid of any conventional treatments or medication but rather by merely relieving them of those excess anxieties that have interfered with some other bodily function. This is why a placebo, such as a sugar tablet, can relieve an individual suffering from a psychosomatic illness. The mere act of believing that a cure has been provided has the effect of lowering that person's anxiety levels. With one's anxiety levels reduced, the ego function can operate more effectively, allowing the nervous system to do the same. Consequently, someone who is psychosomatically paralyzed might regain use of his legs merely by reducing his stress level.

Because the act of prayer has the capacity to reduce stress levels, people who suffer from psychosomatic illnesses are often remedied by this medium. This type of response is most dramatically realized when facilitated by the techniques of those we refer to as Faith Healers.

Take, for instance, the example of the "blind" man at the revival meeting who "miraculously" has his vision restored by the work of a Faith Healer. Often, according to the ailing person, no medical doctor could find an organic cause for his condition and therefore no medical cure could ever provide relief for his disorder. This is because a psychosomatic illness doesn't originate from the allegedly "sick" body part, but instead stems from a debilitated ego function. Though the Faith Healer is incapable of performing miracles, he is adept at tapping into one's prayer function, thus allowing such psychosomatically ill individuals to vent their excess anxieties through the act of prayer. By invoking the prayer function, the Faith Healer is really just facilitating a cerebral catharsis in someone who is stricken by excess anxiety, often related to some childhood trauma or deeply suppressed guilt syndrome. In this way, the Faith Healer works like a placebo. By helping to rouse a psychosomatically ill person's faith in a god that can "heal" and "save," the suffering person is relieved of a great deal of his excess anxieties. Once this is accomplished, the strain that had previously been displaced on the person's nervous system is relieved to the point that the psychosomatically ill person may suddenly find himself "healed." In essence, a Faith Healer helps to relieve a disabled individual of those excess anxieties that have prompted some psychosomatic illness. We could therefore say that the contemporary Faith Healer is serving the role of a modern day shaman, something every primitive society has possessed for exactly this same reason.

What I mean to demonstrate by all of this is that when we are cured or healed through acts of prayer, it is not the result of miracles, but rather the consequence of a purely physiological response to having one's anxiety levels diminished. The fact that all cultures have spoken of the healing properties of prayer leads me to believe that our species possesses a distinct set of prayer-responsive mechanisms that exist within our brains. For this reason, I refer to this capacity as our prayer function.

SIX

<u>RELIGIOUS CONVERSION</u>

"To whom is the Lord revealed?...He is despised and rejected of men, a man of sorrows and acquainted with grief."   — Book of Isaiah

When we speak of a person being "Born Again," we are generally referring to someone who has undergone a religious conversion. When we see someone who has spent his or her life leading a secular existence, suddenly devote him or herself to an organized cult or religion, this is usually the result of a religious conversion. When someone with whom we used to go out to the bars and baseball games is suddenly spending his or her days handing out religious pamphlets in the streets, shouting to every passerby that "God Saves!" this, too, is most likely the result of a religious conversion.

Perhaps we've known someone close to us who has undergone such an abrupt personal transformation or perhaps we have only seen such people as they've proselytized their newfound faiths in the streets. Regardless, the fact that this psychological phenomenon occurs in a cross-section of every population implies that it represents yet another integral facet of our species' inherent natures.

In his book, *Varieties of Religious Experience*, William James was one of the first to document this strictly human phenomenon. As James expressed it, "to say that a man is 'converted' means that religious ideas, peripheral in his consciousness, now take a central place, and that religious aims form the habitual center of his energy."[69]

For people who undergo religious conversions, individuality is replaced by ideology, and very little room is left for personal growth or expression. Because the converted person believes that his or her newfound God and faith determines all things, all personal responsibility is relegated to that religion's leader or supreme being. It is as if the person wishes to revert back to some infantile state in which he is no longer responsible for anything - only now, instead of his parents making all of his decisions for him, the church or God does.

To the converted, all events are determined by God. Individual will is therefore replaced by God's will. To the converted, everything that happens does so because God willed it such. No matter how ordinary or mundane an event may seem, God is perceived as being responsible for everything. Having surrendered all sense of self over to some church, the converted individual places all sense of personal trust in his maker. God is seen as all loving, and all that He does is in our greatest interest. Should something unfavorable occur, it is only because "God works in mysterious ways." Tragic events are now seen as "blessings in disguise." As much as it may seem as if God is punishing us, this is really His way of showing us how much He loves and cares for us. (In many ways, such sentiments mirror those often expressed by neglected or abused children for their abusive parents. As much as the child is mistreated, they regard their abusive parents as caring and loving benefactors.)

Concerning the conversion process itself, though some conversions take place in a slow and gradual manner, a significant number occur very abruptly. Many psychologists such as E.S. Ames favored "restricting the term 'conversion' to sudden instances of religious change."[70] G.A. Coe also thought that the use of the term conversion should be limited to those cases in which the individual undergoes an intense and sudden religious change. The only other example I can think of in which a person's core personality undergoes such an abrupt and drastic change is when one is stricken by psychosis.

And what precipitates these religious conversions? According to the psychologist Paul Johnson, "a genuine religious conversion is the outcome of a crisis...a crisis of ultimate concern...a sense of desperate conflict."[71] In his book, *Religious Conversion: a Bio-psychological Study,* the psychologist S. De Sanctis claimed that "all the converted speak of their crisis, of their efforts, and of their conflicts which they have endured."[72]

In a series of studies done by C. Ullman in *Cognitive and Emotional Antecedents of Religious Conversion,* he compared the converted to the nonconverted. In his studies, Ullman found that "Converts recalled childhoods that were less happy and filled with more anguish than those of nonconverts. The emotions recalled for adolescence followed similar childhood patterns, with the addition of significant anger and fear in adolescence for the converts and not the nonconverts. Converts also differed from the unconverted in having less love and admiration for their fathers, and more indifference and anger towards them."[73] After studying 2,174 cases of religious conversions, psychologist E.T Clark suggested that,"Sudden conversions were associated with fear and anxiety."[74]

From these studies, it seems evident that people who undergo religious conversions often suffered from some distinct emotional instability prior to their personal transformation. Study after study shows that people who are most susceptible to undergoing this particular type of change are those who possess frail senses of identity, people with unhealthily developed, weak, or battered egos.

Follow-up studies show that after such deeply troubled individuals undergo a religious conversion, their emotional states generally improve. According to a study done by J.B. Pratt, "Prior to conversions, individuals had a tendency to wallow in feelings of unworthiness, self-doubt, and depreciation that are released or overcome via conversion."[75] Yet another study showed that, "It is typical of conversion to be preceded by morbid feelings in which doubt, anxiety, internal strife, and despair are replaced by serenity, peace, and optimism."[76] Apparently, for those who suffer severe emotional turmoil, there are obvious benefits in undergoing a religious conversion.

Because a certain percentage of every world culture undergo a religious conversion, such a phenomenon must represent an inherent characteristic of our species, a genetically inherited response to crisis, fear, and anxiety. It seems that religious conversion represents some form of an evolutionary adaptation that serves our species as an emotional support system for those whose spiritual functions alone aren't enough to provide sufficient relief for their excess fears and anxieties. In such cases, one's ego mechanism - one's sense of personal identity - is entirely subjugated, only to be replaced by a back-up or alternate "spiritual" one. Once a religious convert loses all sense of personal identity, he is consequently relieved of all those anxieties attached with personal responsibility. Instead, all such responsibility seems to be relegated to the individual's newfound faith.

The ego is a very delicate organ. If it is not nurtured properly, a person may grow to develop all sorts of insecurities. When a person with a weak or battered sense of self reaches the preliminary stages of adulthood, he may not feel ready or able to take responsibility for himself. Perhaps this is why religious conversions "typically occur during adolescence."[77] "Surveying five studies conducted between 1899 and 1929 on over 15,000 people, Johnson noted that the average age of conversion was 15.2 years."[78]

This is not to suggest that religious conversion only occurs during adolescence, for it can strike at any age that a person feels particularly vulnerable and/or threatened. Nevertheless, it is during adolescence that human beings are generally afflicted with increased anxiety levels, as it is during this time that we are first told by our once-supportive parents that we must learn to fend for ourselves. Adolescence is that time when we are told we must decide how we are going to support ourselves for the rest of our lives, when we are asked to determine the ultimate purpose of our existence. In essence, it is during adolescence that we must first come to terms with the concept of our own mortality.

With all of these concerns, questions, and responsibilities suddenly thrust upon us, it is no surprise that it is during this same stage in our developments that humans suffer the most cases not only of religious conversion, but of suicide, drug abuse, anorexia, and schizophrenia. Obviously, this is a very tumultuous and transitional period in our development.

It is for this reason that I believe nature provided us with this type of a "spiritual" back-up identity by which those with critically vulnerable ego functions can still manage to survive. It is as if nature has equipped us with an emergency replacement identity through which one's dysfunctional ego is replaced by a "spiritual" alternate. It is through this "spiritual" replacement that a person with a particularly vulnerable ego mechanism can at least now function and therefore survive.

Because religious conversion can, at times, save a person from such things as drug abuse, suicide, alcoholism, and a host of other self-destructive behaviors, we tend not to classify it as either a problem or a disorder. It is interesting that when a person converts to a sanctioned religion we are more accepting of it, yet when one joins an unsanctioned one - that which we often refer to as a cult - we feel compelled to seize that person from the "insidious" clutches of such an "unhealthy" group or influence. Regardless of how we choose to perceive this strictly human phenomenon, we must accept that it represents yet another cross-cultural characteristic of our species, one that must therefore represent a genetically inherited proclivity in us.

SEVEN

WHY ARE THERE ATHEISTS?

"What is going to happen to those of us who want to believe but aren't able to? And what is to become of those who neither want to nor are capable of believing?"
— Ingmar Bergmann, The Seventh Seal

In discussing my ideas with others, one of the questions I found myself most frequently asked was, "If human spirituality represents an inherent characteristic of our species, then why are there atheists?"

Genuine atheists are those who do not believe in the existence of a god or a spiritual reality of any kind. Not only do I believe such people exist but that they have probably represented a small percentage of every human population from the dawn of our species. Nevertheless, if the human species is, as I'm suggesting, "wired" to believe in a god, a soul, and an afterlife, then how might one explain the existence of those who don't?

Though we may all exist as a part of the exact same species, no two human beings are exactly alike. As similar as we each might be, we are all unique. As a matter of fact, for each trait we possess we fall into a *bell curve*. For every trait we possess, though the majority of our species will fall into the bulge or the mean of this curve, a small percentage of every population will fall into the curve's opposite extremes.

To demonstrate this notion, let's take a basic physical characteristic such as height. Though the majority of our species is of *average* height, each population possesses a minority of individuals who fall into either extreme of this physical characteristic. For example, though most of us fall into the central bulge of the height bell curve (and are of near average height), on one end of this same curve every population possesses a small percentage of individuals who are excessively short, while on the curve's opposite end, there exists an equally small percentage of individuals who are excessively tall.

As yet another example, let's examine something as basic as eyesight. Though the majority of our species is born with average vision, and will therefore fall toward the center of this curve, there exists a much smaller percentage of individuals within every population who are born with superior eyesight, while on the opposite end of the same curve, there exists an equally small number of individuals who are born totally blind. Again, for every physical trait we possess, though the majority will always fall toward the curve's center, there will always exist those smaller numbers of every population that will fall into one of the two extremes.

Let's apply this same principle to a trait that is more cognitive in nature. Let's take musical ability as our next example. Though most of us are born with an *average* amount of musical ability with which to compose, comprehend, or play music, each population possesses a smaller percentage of individuals who fall on either end of this musical curve. While on one extreme, every culture possesses a small number of those who are born musically gifted (musical prodigies, savants, i.e., Mozart), every culture also possesses a small percentage of those who are born tone deaf or otherwise musically deficient.

For every physical capacity we possess, there must exist a physiological site from which that capacity is generated. Our capacity for vision, for instance, is directly related to the caliber of our eyes and occipital cortex. Similarly, our capacity for music can be directly related to the caliber of that physiological site in the brain from which musical capacity is generated. We could, therefore, say that someone such as Mozart must have been born with an unusually overdeveloped musical part of his brain. On the other hand, someone born tone deaf was born with a rather underdeveloped musical part of theirs.

This is not to suggest that we should exclude the environmental component. Though each of us may be born with a certain degree of potential in most capacities, much of how these inherent potentials are actualized depends on how much we nurture them. Had I, for instance, been provided with a great deal of musical training from early childhood on, I'm sure I would possess a greater degree of musical ability than I do today. Nevertheless, even with the most intensive musical training conceivable, there's no way anyone could have ever made a Mozart out of me.

The same holds true for the opposite scenario. Mozart, were he born to peasants, indentured to toil the fields his entire life, and never afforded the opportunity to study music, would never have reached his full musical potential. In such a case, he may have instead grown to be recognized as "the guy who whistles nicely while toiling the fields." Unfortunately, in this same manner, latent Mozarts, Einsteins, and Michelangelos probably die every day without the slightest recognition simply because they were never afforded the opportunity to actualize their full potentials. I'm therefore suggesting that while life experience plays a significant role in our cognitive developments, we can only reach as high as our inherent genetic potentials permit.

So what does any of this have to do with the question of atheism? Since I am suggesting that spirituality is generated from a specific part or parts of the brain, couldn't we then apply the aforementioned principle to our inherent spiritual proclivities? If we do possess a spiritual function, it would make sense that, while the *average* person from any given population will most probably possess an *average* degree of spiritual consciousness, there exist those smaller minorities who will fall into the extremes of this trait. Regarding those who fall in the mean of this cognitive trait, such people perceive and experience enough of a spiritual reality to believe in the concepts of a god, a soul, and an afterlife. Such individuals will possess a developed enough spiritual function to perceive the existence of a transcendental reality through which they will come to believe that they are immortal, thus enabling them to circumvent their fear of death. These are our masses, the bulge of the spiritual bell curve, those who have kept religion thriving for all these years as an integral part of every world culture. On the extremes of the same curve, however, there exist a much smaller percentage of any given population who are born with an either over or underdeveloped spiritual function.

On the one extreme of this curve lie those born with an overdeveloped spiritual function, those for whom spirituality will play a dominant role in their conscious experience. Such people will tend to be highly spiritual, with a tendency to also be very religious. Such individuals might be found delivering heartfelt sermons at the pulpit from early childhood, those of whom we might say were "born with the spirit in them," or whom we might find "speaking in tongues." Religiously speaking, these are our Mozarts. These are our religious savants, our prophets, zealots, those who are religiously "gifted," born with an overdeveloped spiritual function.

On the opposite extreme of this same curve are those we might call spiritually deficient, those born with an unusually underdeveloped spiritual function. Just as a person born blind would be light-insensitive, those born with an underdeveloped spiritual function are spiritually insensitive, incapable of fully grasping, appreciating, or experiencing the concept of a spiritual reality. Such people rarely, if ever, feel compelled to worship or pray, to consider or contemplate the concepts of a spiritual realm, a god, a soul, or an afterlife. These are the religiously indifferent, the spiritually retarded, if you will. These, in many cases (though not all), are our atheists.

In some cases atheists are those who are inherently spiritual, but because they have become so disillusioned with organized religion have chosen to deny their inherent spiritual inclinations, and have consequently, chosen to deny God. Nevertheless, in many cases, atheists are simply those who, because they were born with an underdeveloped spiritual function are indifferent to spiritual/religious concerns and consequently fall under the heading of non-believers or atheists.

Once again, this is not to exclude the role that environment plays in our spiritual developments. There are those, for example, who, though they might be born with a minimal spiritual function, due to a strong religious upbringing, will still turn out to be highly religious*. At the same time, there are those born with an enhanced spiritual function who, because they were raised in a secular environment, will turn out to be less religious than they might have otherwise been. Like all behavioral traits, the strength of one's spiritual consciousness exists as the result of a combination of one's innate (genetic) potential combined with the influences of one's environment.

*Remember, religion merely represents the cultural manifestation of the spiritual impulse. Therefore, just because someone happens to be particularly religious doesn't mean that they will necessarily be very spiritual. Inversely, someone who happens to be very spiritual might not necessarily be particularly religious. Whereas spirituality, in its purest form, mainly involves a belief in some form of a transcendental reality along with those sensations evoked by such beliefs, religion is more involved with the social organization of the spiritual impulse and revolves more around dogma, ceremony, and ritual. For instance, though a highly religious person might be compulsively dedicated to religious ritual, that same person might not be capable of having a spiritual experience. Regardless, for the mostpart, both of these individual impulses (the one to be religious and the other spiritual), are interlinked in a most essential way.

EIGHT

NEAR DEATH EXPERIENCES

"Mysteries are not necessarily miracles."   - Goethe

We are all familiar, to some degree, with the notion of a near-death experience. Whether we've had such an experience ourselves or have simply heard one recounted by a close friend or on some daytime television talk show, near death experiences constitute an apparent part of the human condition. In reports of those who have survived a car crash or an emergency open-heart surgery, many, while on the verge of death, claim to have experiences through which they are convinced that there must be a life after death. The question is, do near-death experiences offer a genuine glimpse into an actual afterlife or are they merely the consequence of a strictly neurophysiological response to the intense pain and fear evoked by having come so close to death?

Every culture dating back to the beginning of our species has practiced some form of a funerary rite in which, though the body of the deceased is disposed of, it is anticipated that his or her spirit will endure in some next or other plane. One of the chief means by which humans substantiate this belief in an afterlife comes in the form of a particular set of sensations we experience when on the verge of death. Near-death experiences, or NDE's as they are often referred, represent yet another cross-cultural component of the human experience. Due to the cross-cultural nature of this experience, it would appear that this might very well represent another inherent characteristic of our species, that is, yet another genetically inherited trait.

Written references to NDE's date back to *The Tibetan Book of the Dead* and have been catalogued by nearly every world culture since. "In 1982, the Gallup organization published a national survey, 'Adventures in Immortality' which set out to examine what adult Americans believe about life after death. One of the questions asked was, have you yourself ever been at the verge of death which involved any unusual experience?"[9] Fifteen percent of those questioned in this poll said that they had. In China in a similar survey conducted by Dr. Feng and Dr. Lin, forty-two percent of those questioned claimed to have had NDE's.

Though there is no one international standard through which to formally define an NDE, studies show that there exist vast similarities in the descriptions of this phenomenon, ones that cross all cultural boundaries (Fenwick, 1997; Feng and Lin, 1976; Parischa and Stevenson, 1986). For example, in the majority of accounts, those questioned say that the first thing they recall of their experience is a feeling of intense fear suddenly being replaced by a sense of joy, bliss, and peace (similar to more generic spiritual experiences). D.B. Carr suggested (1981, 1989) that such sensations, in so far as they are experienced during a NDE, might come as the result of a flood release of endogenous opiods (endorphins).

The next most often described symptom to occur is that of an OBE or "out-of-body" experience. Here, the person describes a sensation of rising or floating outside of his physical body and, in some cases, even being able to look down at one's self from above*. During this part of the experience, those undergoing an OBE have expressed a sense that their limbs and body are "moving" within their mind, though they are actually immobile. This is similar to the type of hallucinations, or confabulations as they are called, suffered by those who sustain right parietal lesions.

Yet another common symptom of the NDE is described as a sensation of being led down a dark tunnel and then being drawn toward a bright light, one that is often interpreted as having religious significance, such as being drawn toward the gates of heaven. (This description of experiencing a "piercing" or "blinding" white light has been attributed to the flashing of the brain's optic nerve which has a tendency to erratically flare when deprived of a normal supply of oxygen.) It is during this same part of the experience that a person will often express a feeling of being engulfed by God's presence. This would make sense as it seems that intense pain, fear, or anxiety have a tendency to trigger one's spiritual function, similar to the "man in the foxhole" syndrome.

---

*One such hospital in order to validate claims of "out-of-body" experiences, has placed a secret message displayed on an LED marquee above the beds of their patients at a vantage point which can only be seen while looking down from above. So far, not one person who has claimed to have had an NDE or "out-of-body" experience has expressed seeing it.

Following this are descriptions of a sudden, and often painful, sensation of being "slammed" back into one's body, after which the person reports a return to their normal physical experience. Along with these primary symptoms come a number of secondary sensations such as hearing music and smelling scents such as flowers. Symptoms as these can easily be accounted for as simple auditory and olfactory hallucinations.

It is no surprise that a significant number of those who have a NDE not only describe their experience as being spiritual in nature, but also claim that it strengthens their faith in a god. Nevertheless, we must ask ourselves, is this type of experience really transcendental in nature or are we dealing with a concatenation of neurophysical events?

One key to answering this question comes through research that has found that "Near-death experiences can be induced by using the dissociative drug ketamine."[80] Dr. Karl Jansen's report goes on to state that, "It is now clear that NDE's are due to the blockade of brain receptors (drug binding sites) for the neurotransmitter glutamate. These binding sites are called the N-methyl-D-asparate (NMDA) receptors. Conditions which precipitate NDE's (low oxygen, low blood flow, low blood sugar, temporal lobe epilepsy, etc.) have been shown to release a flood of glutamate, over-activating NMDA receptors. Conditions which trigger a glutamate flood may also trigger a flood of ketamine-like brain chemicals, leading to an altered state of consciousness."[81]

It was also found that an intravenous injection of 50-100mg of ketamine reproduces all of the features commonly associated with the near-death experience. (Sputz, 1989; Jansen, 1995, 1996). Even Timothy Leary, the notorious psychedelic drug advocate of the 1960's, described his experiences with ketamine as an "experiment in voluntary death" (Leary, 1983)

Similar to the manner in which entheogenic drugs trigger the symptoms of a "spiritual" experience, the drug ketamine can synthetically trigger the symptoms of a near-death experience. What this suggests is that, as with any other type of spiritual experience, near-death experiences are rooted in our neurochemistry. Apparently, the NDE represents the consequence of a physiological mechanism that enables our species to cope with the overwhelming nature of the death experience.

Once Again, though such evidence can never prove there is no spiritual reality, it is certainly suggestive that this might very well be the case.

NINE

<u>THE GUILT AND MORALITY FUNCTIONS</u>

"Scientists and humanists should consider together the possibility that the time has come for ethics to be removed temporarily from the philosophers and biologized."[82]
                                              - E.O. Wilson

Just as individuals from every culture have possessed the capacity to experience feelings of sadness or joy, every culture has possessed a capacity to experience feelings of guilt - a remorseful awareness of having done something wrong. That all cultures have expressed experiencing this sentiment, and all in such similar terms, suggests that it represents yet another genetically inherited characteristic of our species. We could therefore also presume that there must exist some physiological site or mechanism somewhere in the brain from which this experience is generated. It equally suggests we must possess "guilt" genes which prompt our emerging bodies to develop those neural connections that will come to constitute a guilt mechanism in us.

But what is the origin of such a peculiar sentiment? What exactly is its function? Furthermore, in what way might the sentiment of guilt be related to our spiritual functions?

In order to understand the nature of guilt, we must first chart its evolutionary origins. During the time of the emergence of organic matter, the majority of the Earth's creatures lived independently, as opposed to in groups. This was primarily due to the fact that during those earliest times, all life reproduced asexually and, consequently, had no real need to congregate. In asexual reproduction, one genderless, single-celled organism spawns another by forging an exact duplicate of itself. Due to the nature of this reproductive strategy, there was never any need for any two organisms from the same species to interact.

As life continued to evolve, however, two distinct sexes emerged. Among these new sexually reproducing organisms, it now took two members of the same species, one of each gender, to merge their genes in order to procreate. This new reproductive strategy served to an organism's advantage in that it promoted greater diversity among offspring. Greater diversity meant a greater chance for advantageous adaptations to emerge. The more advantageous adaptations to emerge, the more a species is likely to survive.

Even with the advent of sexual reproduction, the majority of species were still non-social, meaning each individual organism still lived a predominantly solitary existence. The difference now was that the two sexes had to meet at least once in a lifetime in order to procreate. Such gatherings often occurred during a species' mating season in which the two sexes met, usually for the first and only time, merely to copulate and nothing more. Moreover, among such species, once the mother layed her eggs, she usually abandoned them, never to behold her own progeny.

As time went on, and life continued to diversify, an evolutionary trend began to occur in which individual organisms started to live among one another in groups. Within a group, each individual organism was more secure than if it had lived on its own. Within a group, not only could individuals better defend themselves against predators, but they could more effectively hunt and forage. Because of the strength and stability that came with this social adaptation, the group dynamic became the "favored" evolutionary trend, particularly among vertebrates, and most particularly among mammals.

With all the advantages that came with this new group dynamic, there were some disadvantages as well*. In regard to those draw-

---

*There is no such thing as a perfect trait. For every adaptation, as advantageous as it might be, there is always some drawback. For example, though the sickle cell was selected in humans for its ability to help us resist malaria, its emergence constituted its own threat. In this way, evolution works as a seemingly haphazard process of trial and error. As new variations emerge with each individual organism, some are to the individual's advantage, some to its disadvantage, while almost all are are a little of both. In essence, every trait we possess comes with its share of pros and cons. In accordance with the essential physical laws of nature [e.g. the Laws of Thermodynamics], we could say that any given variation to emerge renders an organism either more or less energy efficient. Whereas those variations that happen to be most energy efficient will most likely endure, those least so are most likely to succumb to the forces of extinction.

backs that came with the emergence of the social species: Before the emergence of the group dynamic, individual organisms lived primarily by and for themselves. Because these earliest life forms lived exclusively solitary existences, they did so without regard for any other member of their species. Consequently, all behavior was governed by an animal's self-serving instincts. It was a strictly planarian-eat-planarian world.

As organisms evolved to coexist among one another in groups, these selfish instincts no longer served to an animal's advantage. Obviously, if every creature within a group setting only struggled for its own survival, without any regard for any other within its community, there would be no way for the group dynamic to endure. Now that life forms were evolving to coexist in groups, newer adaptations had to emerge by which a species could balance the needs of the individual with the needs of the group. In other words, organisms had to evolve a capacity to apportion their own needs, so that they could serve themselves while simultaneously serving the needs of the group. Strictly selfish behaviors suddenly represented a threat to the group, and a threat to the group meant a threat to every individual within the group. Though each individual added to a group's strength and therefore served to its advantage, because each individual also possessed its own self-serving instincts, each member simultaneously represented a potential threat.

This was not the only drawback to arise with the emergence of a group dynamic. Now that individual organisms were brought to live in such close proximity to one another, there was an increased likelihood of transmitting contagious diseases. Among the less social species, one single organism infected with a transmittable disease was much more likely to die on its own without infecting any others of its species. Since these social organisms lived in such close contact with one another, now when an individual was infected with a transmittable disease, it was much more likely to infect its entire community.

A third problem of the group dynamic was that it represented a potential threat to a species' gene pool. Since the group worked to protect all of its members, now even the weakest members within the group - within the species - were more likely to survive. On its own, a weak, sickly, or physically handicapped individual is less likely to survive. Among the group dynamic, however, even the weakest members are at least partially protected from any external threat. Consequently, among the social orders, it became more likely that a weaker individual might live long enough to reproduce and therefore to pass its "inferior" genes on to future generations, thus negatively affecting the entire species' gene pool.

Suppose for example, an organism from a non-social species happened to be born with a bad leg or inferior vision. In such cases, not only would that individual find it difficult to hunt and forage for food, but it would have an equally difficult time safeguarding itself against predators. In a group, however, this same physically handicapped individual would have a much better chance of surviving since it would be sheltered by the group. Therefore, though the group dynamic represents a highly advantageous adaptation, it at the same time threatens to compromise a species' gene pool.

When organisms subsist individually, as opposed to in groups, the weakest of the species are more vulnerable and, therefore, less likely to survive. In this way, with every passing generation the weakest members of a species are weeded out (along with their genes) for extinction. Because of this, with every passing generation, each species should be better suited to its physical environment. It should be stronger, more fit [more energy efficient] and therefore more likely to survive.

Among the social species, however, this rule no longer applies. Among such species, the rule becomes survival of the fittest as well as the weakest. Among the social species, the law of survival of the fittest - the principle that determines all natural selection, all organic evolution - is compromised. As a result, the chances of any such species surviving is compromised as well.

As advantageous as the group dynamic may have been, it threatened to interfere with the process of natural selection. Among the social orders, rather than a species' gene pool getting stronger with each passing generation, it now remained stagnant. In order to compensate for these drawbacks, newer adaptations had to emerge among these social organisms.

To circumvent these new obstacles, the social organisms began to develop new mechanisms which enabled them to counter these problems. One such mechanism to emerge took the form of what is called "ostracizing" behaviors. Here, the social species evolved a mechanism that enabled them to distinguish genetically healthy individuals from diseased, handicapped or generally unhealthy ones.*

Once these social life forms had developed a mechanism that enabled them to recognize a disability in another individual, they were then "wired" to be repulsed by such physical deformities, handicaps, or disease. This is manifest in the way that healthy organisms will instinctively shun, avoid, and in some cases even become belligerent toward a weak or sickly member of its own species. Such behavior can easily be observed in the way that mammalian offspring, along with its parents, will tend to shun, torment, and in some cases even kill, the runts of their litters.

Such ostracizing behaviors exist among most mammals and are most pronounced in humans who are, perhaps, the most discriminating creatures of all. Among our own species, such ostracizing behaviors can be most readily witnessed in children as they have yet to be sufficiently socialized to behave more sympathetically towards a mentally or physically handicapped individual.

The purpose of this ostracizing mechanism is to prompt healthy individuals to cast the genetically "inferior" out of their groups and into a state of isolation where they will most likely perish, taking their "inferior" genes with them. This mechanism helped to resolve two of the most essential problems associated with the group dynamic. First, it helped deter the spread of transmittable diseases. Second, this reflex to reject those with substandard genes insured the fortification of the entire species' gene pool.

Even with one threat to the gene pool resolved, there still existed that internal threat to the group generated by those destructive, yet necessary, selfish instincts inherent in each individual within the group. How was nature to balance these conflicting needs of individual self-preservation with the need to preserve the group? Obviously no organism could survive if it lost all of its selfish instincts and concerned itself exclusively with the welfare of others. At the same time,

---

*There are those who hypothesize that the means by which many organisms discern health in another of their own species is through the visual recognition of symmetry in the physical characteristics of that organism. Physical symmetry, it is suggested, correlates to fitness and therefore becomes the mechanism by which we discern a healthy individual from a diseased or handicapped one. In our own species, this same mechanism might be responsible for determining our aesthetic sensibilities by which we call some individuals "beautiful" compared to those we might call "ugly."

no group could survive if each member was exclusively self-serving and never considered the needs of those within its community. For this reason, nature had to select a new mechanism that would balance these two essential yet conflicting needs.

In pre-human social orders, the threat posed to a group by individual selfish behavior was held in check by an evolutionary strategy known as the hierarchy system. In the hierarchy system, every member within a group engages in a series of physical contests until each individual's position in the hierarchy is determined. Whichever individual proves itself strongest will dominate the others as their leader. This dominant individual (often referred to as the alpha male or female) will be first in line to eat when food is procured. More significantly, he or she will also have first choice in the selection of a mate. This will ensure that the fittest male's genes will be coupled with the fittest female's, in turn, insuring the production of the fittest offspring.

Among the pre-human social orders, it was through the hierarchy system that group order was preserved. Despite the fact that the group was comprised of individuals generally driven by their more selfish instincts, the hierarchy system maintained stability and order. In such a dynamic, though a weaker individual might at times be tempted to act on his or her more selfish instincts, such impulses are held in check by the structure of the hierarchy. Should an individual, for instance, try to take more than its fair share of a kill, that individual will inevitably be challenged by one of its superiors. Should such a "greedy" individual decide to dispute its rank or order, it can at any time challenge another member of its group to a physical contest. If the challenger prevails, its position in the group is elevated. If it loses, it will either maintain its old rank or, in some cases, it might even be shunned or chastised by its community for trying to usurp a superior and disrupt the group order. Under the hierarchy system, the group dynamic was maintained by the simple law of domination by the fittest. At no point, for example, could a weaker member claim superiority without eventually being challenged and forced back into submission. In this way, raw physical strength settled all scores and helped to maintain a harmonious order among the pre-human social species.

With the advent of humans, however, this was all changed. Humans, in a sense, represent the end of the physical hierarchy system. Unlike any other species, because of our cerebral capacities, every individual possesses the power to subjugate or to kill any other. Before humans, if a weaker member within a group decided to challenge a superior, he or she would be defeated based on pure physical strength. With the emergence of human intelligence, however, even the physically weakest member of a community possesses the capacity to kill, and, consequently, to displace any other. Among human societies, even the physically weakest member of a community can, should he or she be so inclined, pick up a heavy object and bludgeon the physically strongest member of its community to death. With our enhanced capacity to construct and use tools, the lines of the hierarchy became irrevocably blurred. In light of our intelligence, strength took on a whole new meaning. No longer could a human society rely on raw, physical strength to maintain order. Instead, some newer device was now needed if the group, not to mention the entire species, was to survive. It was at this point in our evolution that a moral function emerged.

Just as all cultures display a distinct set of what we could classify as "spiritual" behaviors, all cultures display a distinct set of what we could classify as "moral" behaviors. Moral behavior can be characterized as that tendency in our species (and only our species) to categorize all acts as being either productive or destructive to the welfare of the group. Those acts viewed as productive to the group are cross-culturally perceived as what we call "good," while those acts we perceive as harmful to the group, we cross-culturally refer to as "bad." This propensity to discern "good" from "bad" behaviors is made evident by the fact that every culture has compiled a list of rules and regulations [laws]. Just as our biological ancestors ostracized those individuals who represented a threat to the group, we do the same to those who break our society's laws.

Though our species may possess some very strong communal instincts, we are still driven, to a significant degree, by our more selfish and destructive impulses. Because of this, it became necessary for our species to evolve a moral function. Just as our ancestors could distinguish a physically healthy individual from an unhealthy one, because our species is so much more behaviorally complex, it became necessary that we develop a capacity to distinguish healthy behaviors from unhealthy ones. Again, those behaviors we perceive as being advantageous to the group, we define as "good," whereas those we perceive as harmful to the group, we define as "bad."

By implementing our linguistic functions, humans possessed the capacity to compile verbal and, eventually, written lists of those behaviors they perceived as being potentially harmful to the group. Once these rules and regulations became codified, the group could threaten to ostracize or punish any individual who transgressed one of its "laws." To enforce these laws, we developed an instinct to punish those who broke them. In essence, humans had evolved a "penal" function to complement our moral one. This penal function represents that cross-cultural impulse in our species to systematically ostracize and/or punish those who transgress our culture's laws. For the majority of our species, fear of such punishment keeps an individual from acting on his or her more selfish instincts. Once we evolved this instinct to enforce our laws, group order could survive despite our more selfish impulses. I imagine that if such a function hadn't emerged in us, the group dynamic, not to mention our entire species, would have most probably succumbed to the forces of anarchy, and with it extinction.

Though our entire species possesses the same language centers in the brain, every culture, based on its own particular historical and environmental circumstance, has developed its own specific language. Though each language may be unique, each contains certain universal characteristics. Likewise, though our entire species possesses the same spiritual impulse, every culture, based on its own particular historical and environmental circumstance, has cultivated its own unique religion. Though each religion may be unique, each contains certain universal characteristics. Similarly, though our entire species possesses the same moral function, every culture, based on its own particular historical and environmental circumstance, has developed its own moral code. Nevertheless, though each culture may possess its own unique moral code, each contains certain universal similarities. For instance, universally proscribed behaviors or taboos, as they are called, exist among every culture. Taboos involving incest and parricide, as examples, represent universally proscribed behaviors. Laws prohibiting the murder of another individual from within one's own group also exist as a universal characteristic of every culture. This is why such accepted axioms as "thou shalt not kill" strike a universal chord in people from every culture. Because we are genetically predisposed to be repulsed by such "anti-social" behaviors, we cross-culturally perceive such codes or laws as universal "truths."

What I'm suggesting is that moral consciousness, just like spiritual consciousness, represents the manifestation of a genetically inherited impulse. Consequently, our species is "wired" to be positively drawn to behaviors that benefit the group, while we are equally "wired" to be repulsed by behaviors that are perceived as destructive, those actions we label as "bad."

Furthermore, it seems that we possess a cross-cultural proclivity to project our spiritual conceptions onto our moral ones. For instance, behaviors that are looked upon as "good" are, in a spiritual context, perceived as what we call "holy," those actions we cross-culturally view as being condoned by our gods. On the other hand, our species is equally inclined to perceive destructive or "bad" acts as being condemned by our gods. Those actions we might otherwise label as "bad" are, in a spiritual context, cross-culturally perceived as what we now refer to as "evil" - a concept for which every known culture has possessed a word. To support this notion, every culture has maintained a belief in "evil" powers or entities (e.g. demons) whose purpose is cause suffering as well as to tempt the fate of our immortal souls. In addition, almost every world culture has conceived of a place where the souls of those who commit "evil" deeds are condemned to suffer eternal damnation. Hell, Niflheim, Tartarus, Gehenna, Jahannan, Bhumis, Karmavacara, and Hades are examples of places we believe "evil" souls are sent after death. On the other hand, souls of the "good" are cross-culturally perceived as being rewarded by the gods. Whether it be Heaven, Nirvana, the Happy Hunting Grounds, Valhalla, or the Elysian Fields, every world culture has perceived a place where "good" souls are rewarded in the afterlife. Again, all of this suggests that our moral cognitions must be integrally interlinked with our spiritual ones. This, in turn, suggests that our concepts of "good" and "evil" must be viewed, like all physiologically generated perceptions, in relative terms, relative to the particular manner in which our species happens to be "wired" to perceive reality.

Even with the emergence of a moral and penal impulse, our species' selfish instincts still tempted us to defy our society's laws. It was here that nature "selected" two newer characteristics meant to help us balance our selfish impulses with the needs of the group.

The first of these new adaptive impulses to emerge in us was that of altruism. In order to balance our selfish impulses, nature installed a device in our species that countered our drive to serve ourselves with one that compelled us to serve others within our community. With the addition of an altruistic impulse, humans were driven to serve others with nearly the same determination that they were driven to serve themselves.

As with any trait, each individual possesses this altruistic impulse in varying degrees. Though the *average* person may possess an *average* proclivity to act altruistically, there exist those individuals who possess either a diminished versus an enhanced propensity to give to others. On one extreme, every culture contains a certain percentage of individuals who are "wired" with an underdeveloped altruistic drive and who are much more motivated by their more self-serving instincts. These would be represented by our society's robber barons, slave traders, power brokers, exploiters, misers, and thieves, people with little or no regard for others within their community and who are only capable of looking out for themselves, those of whom we might say possess no social conscience. For such people, the desire to give does not play a part in their conscious experience. On the other extreme, each culture contains a small percentage of individuals who possess an overdeveloped altruistic drive and who are almost entirely motivated by their impulse to give. These people are most likely to find themselves playing the role of social reformer, missionary, welfare worker, or philanthropist, as examples.

The second trait selected to help us temper our more selfish instincts, I refer to as the guilt function. As I mentioned earlier, individuals from every culture have shown a capacity to experience feelings of guilt, suggesting that a "guilt" mechanism must have emerged in our species to complement our moral and altruistic drives. Whereas our moral and penal functions provide us with a means to discern and then to shun and/or punish others who act on their more selfish instincts, the guilt function provides us with a mechanism that compels us to shun and/or punish our own selves for committing the same selfish acts we find reprehensible in others. Just as our nervous systems prompt us to retreat from such potential hazards as fire, this sentiment of guilt provides us with a mechanism that prompts us to retreat from committing such potentially hazardous social acts as stealing, incest, and murder. Though many self-serving crimes might momentarily serve some primitive self-serving impulse in us, each represents a threat to the group dynamic, which, because we are all a part of the same group, ultimately represents a threat to ourselves. Ironically, it is merely the instinct for self-preservation that dissuades us from harming others.

With the advent of a guilt function, our moral functions were internalized in such a way that we were now "wired" to be just as repulsed by our own selfish acts as we were by those enacted by others. By constantly carrying around these internalized self-critical impulses, each individual was forced to become ever-vigilant over his own selfish instincts.

Just as with all other traits, each individual experiences guilt in varying degrees. Though the *average* person within any given population will most likely possess an *average* capacity to experience guilt, each culture possesses a smaller percentage of individuals who represent the extremes of this sentiment. On the one hand, there exist those individuals born with an underdeveloped guilt function. These people, no matter how much society may try to change them, are incapable of experiencing feelings of remorse. Such individuals represent society's sociopaths, people who are literally incapable of feeling remorse for a criminal deeds, those of whom we might say possess no conscience. Because such individuals possess no internal mechanism that prompts them to contain their selfish impulses, they are often represented by a society's criminal element.

On the other extreme, every culture maintains a cross-section of individuals who possess an overdeveloped guilt function. Such individuals are plagued with excessive feelings of guilt, regardless of whether or not they have done anything wrong. These overly self-critical or guilty individuals may feel a constant need to condemn and/or punish themselves. Such individuals represent a society's martyrs, penitents, and ascetics.

Yet another piece of evidence that supports this notion is the fact that delusions of guilt represent a common sentiment among schizophrenics from all cultures. The fact that such a particular delusion emerges as a cross-cultural symptom of this disorder suggests that the guilt experience must represent an integral part of human cognition.

So what relationship might our guilt function have with our spiritual one? Generally speaking, when we commit a wrongful act, though we might feel guilty toward the victim of our crime, humans possess a cross-cultural propensity to also feel that they have transgressed their gods. This is made manifest by the fact that every culture has conceived of the notion of "sin." When we transgress the laws of our community, we call it a crime. When we transgress what we perceive to be the laws of our gods, we regard it as a sin. The fact that every culture has possessed a word to express this concept suggests that feelings of guilt have a distinct tendency to evoke one's spiritual consciousness.

To further support this notion that our guilt function is integrally linked with our spiritual one, all cultures have maintained rites through which we seek to repent or "atone" for our sins. Such penitent behaviors are clearly related to the sentiment of guilt.

When the majority of our species commits a wrongful act, it seems to evoke a great deal of anxiety. Much of this anxiety can be attributed to the fear of social and/or divine retribution. Moreover, anxieties evoked by guilt have a tendency to stimulate spiritual consciousness, which often has the effect of turning men to God (This may help to explain, for instance, why prisons often contain such an abundance of religious converts.)

> "Moral anxiety based on guilt and guilt feelings activates religious concerns...In fact, the existence of morality is, to many people, impossible without established religion and belief in God."[83]

But is morality truly contingent on one's believing in an established god or religion? For instance, should all atheists be viewed as inherently immoral? Though they are often stigmatized as such, I contend this to be an unfounded bias. Physiologically speaking, though an atheist might not be innately "hard-wired" to possess a strong spiritual consciousness, his moral centers might be more devoloped than an overtly religious and/or spiritual person. Again, we are talking about at least two distinct modes of consciousness, two types of "wiring," one moral, one spiritual - two centers that might be as unique to one another as are our faculties for language and music. It's therefore just as likely that we could find a sociopathic atheist as someone sociopathic and religious (take, for instance, those from any given inquisition). Consequently, religion and morality should not be viewed as any more synonymous than should atheism and immorality. To counter this stigma, some atheists refer to themselves as "secular humanists" to define their sense of moral responsibility.

TEN

THE LOGIC OF GOD: A NEW PARADIGM

"We are what we think.
All that we are arises with our thoughts.
With our thoughts we make the world." — Buddha

"Projection makes perception. The world you see is what you gave it, nothing more than that....It is the witness to your state of mind, the outward picture of an inward condition. As a man thinketh so does he perceive. Therefore, seek not to change the world, but choose to change your mind about the world." — Anonymous

"An evolution of consciousness is the central evolution of terrestrial existence...a change of consciousness is the major fact of the next evolutionary transformation."[84]
— S. Aurobindo

So, what if Kant was right? What if all of our conceptions of reality are nothing more than the products of internally generated cognitions, "the outward picture of an inward condition?" In such a light, we must accept that all we interpret as knowledge or truth is subjective, relative to the manner in which our species is "wired" to perceive the world.

Because each species acquires as well as processes information differently, each species consequently interprets reality from its own unique and relative perspective. As all of our perspectives are relative, no species, nor any individual within a given species, has access to anything that would constitute absolute knowledge. In those wonderful Kantian terms, we can never possess knowledge of "things-in-themselves," but only of "things-as-we-perceive-them." Just as flies possesses fly knowledge, humans possess human knowledge. And just as flies possess fly "truths," humans possess human "truths," neither one being representative of anything absolute.

Our perspectives of reality are framed by the manner in which our brains process information. Therefore, to better understand the nature of *our* realities, we first need to understand the nature of how our brains function.

The human brain consists of an interactive network of separate and individual mechanisms or processors often referred to as cognitive schemas or functions. Therefore, in order to better understand the nature of how we process information, we first need to understand the nature of each of these interactive cognitive functions.

Examples of human cognitive functions include a language function, a mating function, a musical function, a maternal function, a hygiene function, a mathematical function, an ego function, a variety of emotional functions. As a matter of fact, for every cross-cultural behavior our species exhibits, it is likely we possess a cognitive function from which that behavior is generated.

It is the role of each of these separate cognitive functions to process data, each in its own particular way. Only after all of this separately processed data has been integrated are we provided with a comprehensible - that is, functional - perspective of the world.

So what if we were to apply this same precept to human spirituality? What if spirituality represents the manifestation of one of our brain's cognitive functions? As all cultures perceive reality from a spiritual perspective, is it not possible that spirituality may represent one of the ways our species is "wired" to processes information and consequently to interpret reality? If so, it would imply that our cross-cultural beliefs in such concepts as a god, a soul, and an afterlife constitute nothing more than the products of inherited perceptions, manifestations of the particular way our species happens to process information and, therefore, to interpret reality. In such a light, God would no longer represent an absolute truth, but rather a cognitively generated, relative perception. In essence, such a hypothesis would imply that God, as we've thus far interpreted him (as a real and absolute entity), is, as Nietzsche suggested, dead. No longer an absolute reality, God is reduced to just another one of our species' relative perceptions, the manifestation of an evolutionary adaptation, a coping mechanism installed in us to enable our species to survive our unique and otherwise debilitating awareness of death.

I realize that it may be difficult for many people to accept such a physical/mechanistic/organic/evolutionary/cognitive/rational, that is, scientific interpretation of God. Because the majority of our species is physiologically "hard-wired" to perceive a spiritual reality, it may be impossible for many to grasp this concept, as it may conflict with their inherent perception of reality. As a matter of fact, trying to convince someone who is "hard-wired" to believe in a spiritual reality that no such thing exists may be as futile as trying to convince a schizophrenic that the pink elephant he sees prancing about the room is nothing more than the product of his perceptions. This is not to suggest that our spiritual perceptions represent the manifestation of any physical dysfunction, as is true of the schizophrenic. On the contrary, what I'm suggesting is that spiritual consciousness represents a normal and essential part of human cognition.

But what if we could somehow get the schizophrenic to recognize that his visual hallucinations were nothing more than the products of erroneous perceptions? What if we could somehow teach him to reason through his delusions? Similarly, what if our entire species could come to recognize and to accept that our beliefs in a spiritual realm, a god, a soul, and an afterlife aren't representative of any actual "transcendental" reality but are rather the manifestations of internally generated misperceptions? Furthermore, what if we could learn to see through these inherited delusions? What if we could recognize that all of our spiritual beliefs exist as nothing more than the consequence of a neurophysiological reflex? Just as all planarians turn towards light, humankind turns to God. Just as all peacocks display their feathers when exposed to an aroused peahen, humans believe that, by virtue of their "souls," death is not final.

Imagine an android (a machine built to look and act exactly like a human) is programmed to believe it is human. Imagine that in order to make this android believe such a thing, the manufacturer installed a computer chip into its circuitry that instilled it with fictitious memories of a fabricated past.* Now imagine that the android were to suddenly become aware of its true nature. Suddenly, it realizes not only that it's an android, but that its memories are nothing more than the effects of a computer chip, one that compels it to perceive a delusional past. Now that the android has become cognizant of the true nature of its situation, it would be free to explore the possibilities of

---

*Similar to the plot of the film, *Bladerunner*.

a whole new paradigm. No longer bound to the same false reality with which it was preprogrammed, the android would now be able to redefine its own destiny, to explore new possibilities in accordance with its true nature.

Analagously, imagine humans were to suddenly become cognizant of the fact that we've been programmed by the forces of natural selection to perceive a spiritual reality, one which is as fabricated as the android's fictitious memories. Just as the android had been constructed with computer chips that frame its thoughts and actions, humans are constructed with cognitive functions that do the same for ours. What if, in the same way that our android recognized the fact that its memories existed not as the recollection of actual experiences but rather as the consequence of a program installed into its circuitry, we came to recognize that spiritual consciousness exists, not as the effect of any actual transcendental reality, but rather as the consequence of a neurophysiological mechanism installed into our species' organic hardware? Perhaps if we learned to regard human spirituality in such a way, we, too, could devise a whole new paradigm for ourselves, one through which we could redefine our own destinies based on our "truer" natures. Rather than having to be stuck in the same delusional framework nature forged for us, we could use this self-knowledge to reach toward something better, something through which we could perhaps make ourselves more survivable, more energy efficient. Should we choose to embrace such a new "spiritual" paradigm - one based in scientific reason - it might mark the advent of a new stage in our species' evolution.

As another metaphor, imagine we are looking into a mirror that can offer us a pure and perfect reflection of ourselves. Now imagine that placed between us and this perfect reflection is a series of invisible lenses, one that will distort our otherwise perfect view of reality in some way. Because we are ignorant that these lenses exist, we have no way of knowing that our perceptions of ourselves are being distorted. Though we may believe that the mirror is offering us a perfect view of ourselves, we are misinformed. Until we know these lenses exist, until we know to look past them, to push them away, we will never be afforded a true reflection of ourselves.

I believe that human spirituality represents such a lens, one that distorts our view of reality by making us perceive a spiritual element when, in actuality, no such thing exists. What if we were to become cognizant that this lens existed? What if we were to choose to push it aside, clearing our view of all such "spiritual" distortions, affording ourselves a much clearer, less obstructed view of reality? Sure, it might be somewhat uncomfortable at first, even distressing, to have to readjust our perceptions of ourselves in such a fundamental way. But wouldn't we prefer to possess a more perfect view of reality than a distorted one? Shouldn't we prefer truth over deception?

What I'm suggesting is that spiritual consciousness represents nature's white lie, an inherited misperception selected into our species, for the purpose of alleviating us of some of the anxiety caused by our awareness of death. And why would nature lie to us in such a way? Nature isn't concerned with such lofty notions as what is "real" or "true." Nature is simply concerned with creating a more survivable organism, this and nothing more.

As terrifying as the prospect of death might be, if such an organic theory of spirit and God happens to be correct, isn't it in our best interests to embrace the reality of our situation? Can anything really be gained by living in conscious denial of the truth? As is the case with any white lie, we are going to have to ask ourselves: would we prefer to live in contented ignorance or would we rather embrace the truth? How we choose to answer this, may very well represent one of the pivotal issues underlying the future of our species.

ELEVEN

WHAT, IF ANYTHING, IS TO BE GAINED FROM A SCIENTIFIC INTERPRETATION OF HUMAN SPIRITUALITY AND GOD?

"Religion is the source of all imaginable follies and disturbances. It is the parent of fanaticism and civil discord. It is the enemy of mankind."
- Voltaire

"Science is the great antidote to the poison of superstition. An ailing world would do well to reach for the right bottle in the medicine cabinet."
- Adam Smith

Suppose, for the moment, that what I'm suggesting is ludicrous, the scribblings of a frustrated atheist. Suppose there really is a spiritual realm, a creator, a soul, and an afterlife. Suppose that "we," the essence of our conscious selves will persevere for all eternity. Should this be the case, humankind is free from the threat of death. If we truly are immortal, these bodies we presently inhabit constitute nothing more than superficial skins, which, once shed, will be replaced by another, or perhaps better yet, not replaced by anything at all, spirits set loose to explore the cosmos eternally free from the burdens of any restrictive physical reality. Regardless of what particular circumstance eternal life might bring, as long as God exists, as long as there is some supreme transcendental force that has endowed us with an immortal soul, humankind is saved.

Presuming then that God does exist, what harm can there be in merely considering the possibility that He does not? If God exists, what's there to lose in pondering His potential non-existence? If nothing else, why not simply indulge ourselves in a little mental masturbation while we pass some of our endless time?

With this now in mind, regardless of what the truth is, let's investigate the possibility that God does *not* exist. Let's presume that God and spirit are merely physiologically generated misperceptions, cognitive phantoms instilled within our brain. If this is the case, what might this mean for us as a species? What are the implications of existing in a godless and spiritless universe? Without God, how are we to gauge our conduct? How are we to define morality? Where are we to turn to find purpose or meaning in our lives? Without God, is all necessarily lost, are we truly that defeated, or is it possible that we might find meaning and purpose in other ways? Is it possible that we might even be able to use this newfound understanding of ourselves as a means to improve upon our existences? In essence, what, if anything, might be gained from a scientific interpretation of human spirituality and God?

In order to answer this question, we must first ask: what is it that we might want to gain? What do we, as individuals as well as a species, want out of life? Is it possible that such a universal goal as this even exists? More importantly, if such a universal goal does exist, is its fulfillment contingent on God's existence?

So, is there any one thing that every member of our species universally wishes to achieve in life? At the recommendation of one of the greatest thinkers in human history, I'm going to suggest that such a universal goal does exist. As suggested by Aristotle over two thousand years ago, all humans strive to achieve one thing before all else and that is the greatest amount of happiness in life. This, he postulated, constitutes humankind's "*Summum Bonum*" - its greatest good. According to Aristotle, all we do is done in the hope that it will bring us greater happiness (or, as we might say, minimize our pain and suffering). This, I would agree, represents the universal end of all human action. Furthermore, it would seem that this principle holds true regardless of whether or not a god exists. After all, under what circumstance might humans ever seek to be less happy? We can therefore say that whether God exists or not, our ultimate goal is still the same. Without a god, all is not necessarily lost.

Presuming that maximizing happiness represents the desired end of all human action, how are we to reach this end, most particularly in a potentially godless universe? Just as the acquisition of happiness might represent the universal end of all action, might there be some universal means by which we might achieve this universal goal? Since this brief stay here on Earth might represent our one and only shot at existence, it would seem all the more critical that we be able to answer this question within our lifetime.

In seeking a universal key to happiness, I find myself once again drawn to one of the great ancients. As much as they may have disagreed with one another, almost all of the world's philosophers have concurred in that the key to happiness lies in the acquisition of knowledge. And of all the various forms of knowledge, the greatest, we are told, lies in self-knowledge. Before all else, said Socrates, "*Gnothi Seauton*" - Know Thyself.

It is only because our species possesses a cognitive capacity for self-awareness that humans can even aspire to acquire self-knowledge. No organism from any other species possesses our capacity for self-conscious awareness. Consequently, no other organism can recognize its own individual shortcomings. Because we *can* recognize our own shortcomings, both as individuals as well as a species, unlike any other animal, humans possess the unique capacity to adapt themselves in such a way that we can turn a shortcoming into a strength. In other words, we have the capacity to fortify ourselves, to make ourselves more survivable, more fit. The more survivable we are, the more secure we feel in the world. The more secure we feel, the less anxious we are; the less anxious, the happier we will be. In this way, humans possess the unique capacity to adapt themselves in such a way that we can make ourselves happier.

For example, a man recognizes that he is physically weaker than his peers. To compensate for this shortcoming, he can do any number of things from lifting weights to increase his physical strength to developing some other capacity, such as his intellect, as a means to more effectively compete with his peers. The more effectively a person can compete with his peers, the more secure one feels; the more secure, the happier.

As another example, a man finds himself lonely and consequently unhappy. After pondering his circumstance, this same man comes to the realization that much of his loneliness exists as the result of his excessively selfish tendencies, something that has driven away most of his family and friends. Recognizing that his selfish ways represent the chief cause of his loneliness and consequent unhappiness, he can now use this "self-knowledge" to transform his circumstance. He might, for instance, use his newfound self-knowledge to contain his selfish instincts and to be more giving and considerate. Perhaps, in time, this same man may find himself with more friends and consequently happier. Again, only humans possess this power of self-transformation. As a matter of fact, it constitutes one of, if not the most, significant advantages of self-conscious awareness.

We can also apply this principle of self-modification to the survivability and consequent happiness of our species. For example, should the Earth undergo another ice age, human beings will be able to transform their environments with a versatility shared by no other creature. As suggested earlier, rather than having to wait millions of years for natural selection to provide us with a thicker coat of fur, we can sew ourselves one within a few hours time. As a result, humans can make themselves more survivable, thus more secure, and consequently happier. Once again, knowledge is power, self-knowledge being perhaps the most potent knowledge of all.

If we are to accept this merger of Aristotelian and Socratic precepts then we agree that the universal means of maximizing happiness lies in knowing as much as we possibly can about ourselves. Moreover, if a great deal of our behavior is guided by genetically inherited impulses, then in order to maximize our capacity for happiness, we must first seek to maximize our understanding of those cognitive processes from which all such impulses are generated. Being that certain human impulses can be destructive in nature, by learning to understand them, we will be better equipped to contain them. Granted, because our impulses represent such an integral part of our biological make-ups, it is unreasonable to expect that they can ever be completely suppressed. Nevertheless, by understanding the underlying nature of our destructive impulses, we can at least try to channel those same energies into more productive outlets.

So what if it should turn out that human spirituality is nothing more than the consequence of an inherited biological impulse? If this might be the case, shouldn't we at least inquire into the nature of such an essential part of us? As stated previously, no trait is perfect. Though each physical characteristic we possess provides us with some utility, each comes with its own drawbacks. Consequently, if spirituality constitutes an inherent physical characteristic, what might its drawbacks be? What negative impact might a spiritual function have on our species? Only once we determine this, will we be able to maximize on this impulses' positive aspects while, at the same time, minimizing its negative. Only once we view spiritual consciousness as an evolutionary adaptation will we be able to objectively examine its potential drawbacks. Only once we identify such drawbacks, can we begin to work on turning them into strengths.

For all the advantages there might be in possessing a spiritual impulse, for all the comfort it might bring, it has proven itself, time and time again, to be a very hazardous instinct in us. As the philosopher Alfred North Whitehead expressed it,

> "History, down to the present day, is a melancholy record of the horrors which can attend religion: human sacrifice, and in particular the slaughter of children, cannibalism, sensual orgies, abject superstition, hatred as between races, the maintenance of degrading customs, hysteria, bigotry, can all be laid at its charge. Religion is the last refuge of human savagery."[85]

Granted, humankind has done away with many of the more primal excesses of its spiritual impulse. For example, none of the world's present-day mainstream religions incorporate such rites as child sacrifice or cannibalism into their practices. Nevertheless, even with the proscription of these more barbaric rites, religion continues to act as a divisive force, promoting discrimination among peoples as well as inciting enmity, aggression, and ultimately war.

But why is it that the world's various religions, whose tenets are so often based on just and loving principles, so frequently find themselves pitted against one another in acts of cruelty, hostility, and genocide? Though every culture possesses the same inherent spiritual impulse, because each culture emerges from its own unique background, this same impulse is manifest differently in each culture. It is for this reason that, though we all possess the same inherent "spiritual" impulse, so many different religions have emerged.

Because each religion has faith that its belief system - and only its belief system - is representative of "the truth," the beliefs and ideologies of each religion inherently contradict one another (If my God is true, how can yours be also?). Consequently, there exists an inherent antagonism in each religion for the next. Moreover, our spiritual function instills us with a belief that we are immortal. Because each religion possesses its own interpretation of a soul and an afterlife, each religion inherently perceives every other as a direct threat to its sense of immortality. As a result of this psychodynamic, our species tends to engage in what could be termed *spiritual* or *religious tribalism*. It is this same dynamic that has compelled us to initiate the types of religious wars that have plagued our species' history.

Perhaps if we could learn to view spirituality as nothing more than a genetically inherited impulse, we would be better equipped to contain some of its destructive influences. If we could learn to understand the underlying nature of our spiritual instincts, perhaps we'd be better able to temper the inevitable antagonism that each religion inherently feels for every other. If we were to recognize that our spiritually generated fears and antipathies were merely the effects of an inherited impulse - as opposed to anything founded in reason - we could learn to curb this same impulse that has launched our species into a history of repeated religious war. How many more times must we justify acts of cruelty, murder, and genocide in the name of God and religion before we learn to tame this destructive impulse in us? Even today, we need just look to the Middle East, Northern Ireland, and Yugoslavia, as a mere few examples, to witness the destructive grip the spiritual impulse apparently has on our species. Perhaps if we were to accept a more organic interpretation of our spiritual natures, we could learn to deter such future wars and hatreds. Rather than simply justifying acts of violence in the name of God, we might learn to recognize that we are being driven to commit such atrocities for no other reason than that we have this hazardous impulse implanted in our head. After all, only by curbing our self-destructive impulses will we be able to minimize life's suffering and maximize our capacity to procure the greatest amount of happiness in life.

During the time of our species' emergence, when humans lived in small nomadic communities, perhaps it was necessary that we possess a spiritual impulse. At that time, spiritual consciousness provided us, not just with a means to cope with anxiety and death, but also with a means to order ourselves socially. But times have changed since then. Since the dawn of our species, not only have humans successfully populated the planet but, in the process, we have evolved from close-knit, communally nomadic creatures into citizens of diverse and stationary civilizations. The problem is that these civilizations are very different from the physical environments into which we were originally adapted. Within a relatively short period of time, humans had transformed their own environments into something very different from those for which they were originally selected. As a result, we were no longer biologically suited to meet the demands of these new conditions. At the time of our emergence, we were little more than what Desmond Morris referred to as "Naked Apes," monkey people who lived in caves, could start fires, and chip rocks. And look at us now, a mere hundred thousand years later (which is very little in terms of evolutionary time) with our concrete megalopolises, and advanced modes of communication and transportation which render tribal isolationism an obsolete concept. In essence, the environment for which we were initially "hard-wired" had been drastically modified within a relatively short time span. As a result, certain aspects of our initial "hard-wiring" no longer suited our new conditions. Consequently, we had become an environmentally maladjusted species.

In light of these changes which had taken place, spiritually tribalistic behaviors came to represent a potential threat to the fabric of our new social arrangements. Consequently, humans have had to curb their most primal spiritual impulses. Because we possess these now inappropriate spiritual proclivities, a new set of conflicts have arisen within these civilizations. And so, humans find themselves confronted with the dilemma of having to either abandon their new social institutions or to force an accelerated *social evolution* upon themselves. In order to accomplish this, humans will have to learn to temper their now obsolescent spiritual impulses. As we find ourselves living in what is becoming an ever-increasingly global society, maintaining a diversity of belief systems may no longer represent a functional option for our species. As a result, it is quite possible that religion represents an evolutionary dead-end, one that is destined to become the next vestige of our species.

This notion of containing our more self-destructive impulses seems particularly relevant today in a world in which we possess modern weapons of mass destruction. In such potentially dangerous times, can we really afford to leave ourselves in the hands of our most primal instincts? Just as we need to contain the excesses of all our inherent impulses, shouldn't we seek to do the same for our spiritual ones as well? Rather than to simply learn new ways to negotiate war, wouldn't we be better off if we sought to understand and then contain those impulses that continue to drive us toward one. There is no time left to negotiate. We've played our last chip in the war room. Any next world war that might emulate those of our past would mark the end of life as we know it. As Einstein so eloquently expressed it, "I do not know with what weapons we will fight World War III, but

World War IVwill be fought with sticks and stones."

Just because we as a species are temporarily king of the hill, we presume we're invincible. It's as if we've placed unconditional trust in some force to preserve us, as if we believe that because of the great strength we possess we are immune to the forces of extinction. Perhaps we feel this way because we continue to believe in the myth that we are God's "chosen creatures." To recognize what a puerile fantasy such thinking represents, we need just look at terrestrial life's three and a half billion year history to see that it is little more than a chronicle of mass extinctions. As a matter of fact, for every species that exists today, there are countless numbers that have been weeded out. For us to presume that we are immune to such a fate may very well represent the beginning of the end of our species.

Furthermore, just because we happen to live in a time of relative peace and calm, we shouldn't presume that things will remain this way forever. The history of our species is an epic of war, one that is often contingent on the world's economic conditions which happen to be cyclical in nature, fluctuating between periods of growth and recession. In a period of growth, we become complacent. In recession, we go to war.

In addition, with all of our newly advanced medical technologies, which decrease infant mortality rates and extend life expectancy, the continued rise in our world's population only exacerbates the possibility of a world recession. Again, in light of the destructive technologies we now possess - nuclear, chemical, and biological - perhaps it's time we start to work on mastering our own natures, most particularly on those instincts that tend to be the cause of so much war and conflict. Once we come to understand the nature of our self-destructive instincts only then will we be able to channel these same energies into more productive outlets, and, consequently, to maximize our species' potential for survivability and happiness.

Furthermore, because our spiritual functions compel us to believe in an afterlife, it's possible that we take less responsibility for our actions. Because we instinctively believe that this life represents a mere stop-over in eternal existence, we allow ourselves to be profligate. Because we inherently perceive ourselves as being immortal, we place less meaning and significance on perfecting ourselves within this lifetime as well as in preserving the conditions of this, our Earthly environment. After all, why worry about the Earth when we'll be spending the rest of eternity somewhere else? How else could we possibly justify the manner in which we continue exploit and butcher our planet?

So why shouldn't we choose to use the same methodology that brought us computers, heart transplants, and space ships to make the most of ourselves as well? Isn't it time we began placing the same emphasis that we do in perfecting our toys into perfecting uorselves? Why do we find it so difficult to pass the torch of faith from destructive religious credos to science? Why are we so afraid to let go of the antiquated paradigms by which we were raised? What if our great, great, great, great, and then some, grandparents were wrong? What if those who viewed lightning as God's wrath were mistaken? What if their primitive interpretations of how things work were wrong? Moreover, what harm could there be in at least experimenting with the tools of science as a means to fortify ourselves, as a means to minimize the pain and maximize the happiness? As a matter of fact, the maximization of human happiness may very well represent our best, if not last, shot at survival.

So, which will it be? Are we to accept the underlying principles conceived in scientific method - in reason - or are we to obstinately hold on to those antiquated belief systems that sprang from our prescientific, ignorant past? In the past, it was considered a mortal sin to believe that the Earth revolved around the sun. Since such primitive times, science has sent men to the moon and back. In the past, it was considered a sin to perform an autopsy, that is, to study human anatomy and physiology. Now, as a result of scientific method, we have developed a vast number of medical technologies which have significantly extended the human life span. And yet, in a society as modern as ours, in the world's most powerful democracy, we still find ourselves battling against the suppressive forces of religious conservatism. We still live in a nation in which the same evolutionary principles that brought us so many life enriching technologies struggle to be taught in the classroom. The threat of ultraconservatism still looms over our great society. And why? Because religious values continue to play a significant role in human nature and therefore in human politics.

We rely on our religions to tell us what is acceptable versus unacceptable, what we should and shouldn't do, what we can and can't think. In this way, religion acts as a constricting force, constantly trying to impede the flow of information that might be construed as a threat to its own obsolete ideology. In this way, religion confines us. It limits our field of vision. It tries to place us in a narrow box and bind us within. Should we seek to step out of that box, to merely take a peek at the world of possibilities, we are to be shunned and punished. Why then, when this life may be our last, should we want to limit ourselves in such a way?

Not to suggest that there should be no limits set on human behavior. As a social animal, with often runaway impulses, there's nothing wrong with a bit of healthy restraint. By no means am I encouraging the dissolution of all codes of behavior and conduct. It's just that do we necessarily want such codes, such guidelines to be based on obsolete mythologies? Through the careful application of scientific method, we know more about the nature of human behavior than ever before. Why then, would we want to rely on systems that were based on the whims of men's imaginations, on untested and unproven hunches? For example, if one suffers from psychosis, should they seek out the care of a licensed psychiatrist or an exorcist? Isn't it time we finally discard our antiquated paradigms, and replace them with methods that can at least be validated? How much more evidence do we need before we finally embrace the scientific process? And if we do, shouldn't we seek to resolve our social and ethical dilemmas through this same medium? As I've stated, it's time we seek out newer and more effective paradigms through which to maximize our species potential for survival and happiness.

Suppose, there is no spiritual reality. Suppose we are nothing more than strictly physical entities, a chance combination of molecules, devoid of any ultimate purpose or meaning. If this is the case, then death truly does represent the end of individual consciousness. Granted, energy can neither be created nor destroyed. Granted, the same energy of which we are composed today will exist in some form until the end of time. Nevertheless, once our brain dies, once its cognitive processes stop functioning, so does consciousness. In whatever form our present store of energy will be redistributed into the vast universe after physical death, whether it be as soil, gas, or cosmic dust, it will bear no relation to what we are today. Never again will we exist in the same exact molecular combination. Consequently, never again will we experience the same conscious state.

Since this may represent our one and only shot at existence, shouldn't we place our priorities and emphasis on our lives here on earth rather than on some dubious hereafter? We should revel in life, all life. We should honor organic matter, devote our species' energies to its preservation. As the strongest and most capable of earth's creatures, we must take the helm as the guardians of this planet. As the only species with the power to destroy the world with the push of a button, it's imperative that we assume this role. We have but two choices: we can either work to preserve life or to destroy it. We know we have the capacity to destroy it. The question is, do we have the power to preserve it?

Suppose we are composed of matter and nothing more. If true, we must learn to view ourselves as organic machines. Not until we accomplish this will we be able to effectively act as our own mechanics. If it's true that humans do possess a spiritual function, one that has instilled our species with an impulse leading us to acts of aggression, hostility, and war, shouldn't we seek to master it? If we truly are ticking biological timebombs, shouldn't we seek to diffuse ourselves?

Besides, if there truly is no spiritual reality, just think of all of the energy we've wasted in practicing our illusionary beliefs. Think of all of the useless rituals and ceremonies we've performed, all of the sacrifices we've made, the shrines we've built, the purses we've filled, the gods to whom we've worshipped and prayed and, meanwhile, all of it in vain. If there truly is no spiritual realm, we've been little more than a species of primitives paying homage to thin air. Imagine what a group of onlooking extraterrestrials would think after witnessing our behavior. "Look at the monkey people," they would say, "offering sacrifices to the void; killing, defiling, and warring with one another over literally nothing; banging their chests and wailing at the wind, all in the vain hope that it might incite some non-existent force or being to save them from their inevitable fates."

If it's true that there is no spiritual reality, God, soul, or afterlife, then let's accept ourselves for what we are and make the most of it. Perhaps such a change in our self-perceptions might help us to shift our priorities from the hereafter to the here and now, to deter future wars, and, ultimately, to minimize pain and suffering and to maximize our chances of obtaining the greatest amount of happiness in life. This, more than anything, is what I would hope to gain from a scientific interpretation of human spirituality and God.

EPILOGUE

QUEST'S END

Here lies the end of my personal lifelong quest for knowledge of God. Though I'll always remain open to the possibility that a spiritual/transcendental realm or force might exist, in the meantime, I trust in, that is, I have faith in the solution I've provided for myself.

Granted, I would have preferred if my research yielded proof of a God, proof that there existed some transcendental realm through which "*I*," my conscious self, would live forever. Sure, I would have preferred eternal existence over inevitable death...Then again, would I? How certain am I that I would have wanted to be immortal, to exist forever? Perhaps in the end, it's actually better this way. Imagine for a moment the ramifications of eternal life, of knowing that no matter how exhausted one might become there will never be a moment's rest or respite from existence.

Besides, amid eternity, what goals could one have? How relevant would anything be? Eventually, hours, years, eons would all blur together, rendering existence an endeavor in obscurity. It would be like being in a race with no finish line. Under such circumstances, I can only imagine that, after a while, one would eventually begin to slow down, to stop pushing oneself to achieve. In such a light, what would achievement even mean? Perhaps it's better this way, better to burn quick and bright than forever dim. Perhaps without death, life has no meaning. Perhaps so, perhaps not. Perhaps I'm simply trying to rationalize my fear of my inevitable demise.

So where to now? Knowing that I'm destined to grow old and die, to lose everything I ever had or loved, including my self, "Why," I sometimes ask, "should I bother to continue living?" Why, in such a godless universe, should I continue to push this burdenous rock of Sisyphus for just one more day? Why not just get it over with and kill myself right here and now? Though during some of the more painful and distressing times in my life, I might be enticed by such notions, I console myself with the realization that if there really is no spiritual realm, no God, no soul, no afterlife, then I'll have all eternity to *not* exist, to *not* have to endure the vagaries of capricious reality. With this in mind, why not make the most of this fleeting experience called life while it's still available to me? Even if I were to procure just one more moment of genuine happiness that would still be one more than nothing.

Perhaps the mere fact that we are capable of experiencing anything at all is reason enough to celebrate. After all, how many other combinations of matter can do this? Even if it should turn out that we are just spiritless atoms cavorting in the void, we are still matter's paramount form, the height of its complexity, its creme de la creme. What other combination of atoms possesses the capacity to love; to laugh; to reason or philosophize; to appreciate works of art, literature, and music; to contemplate its own existence; to aspire to be something more than it actually is; to dream; to hope?

Even if what we call happiness is nothing more than the manifestation of a physiological process, do I experience it any less? Whether I'm mortal or immortal, a spiritual entity or a spiritless organic machine, are these not my experiences? Either way, am *I* any less *me*? Besides, the mere fact that I can never know what each next moment will bring means that, as mechanical as life might be, mine will always remain a beautiful and wondrous mystery.

– THE END –

ADDENDUM:

EXPERIMENTS THAT MIGHT HELP PROVE THE EXISTENCE OF A SPIRITUAL FUNCTION

"Scientific method today has reached about as far in its understanding of the human mind as it had in the understanding of electricity by the time of Galvani and Ampere. The Faradays and the Clerk Maxwells of psychology are still to come; new tools of investigation, we can be sure, are still to be discovered before we can penetrate much further, just as the invention of the telescope and calculus were necessary precursors of Newton's great generalizations in mechanics."   — Julian Huxley

"The truth will out!"   — Shakespeare

Though the scientific community has just begun to look into the possibility that there might exist a genetic precursor to human spirituality, we are still in the incipient stages of conducting research and acquiring hard evidence that might support such a hypothesis. The following are experiments I believe could help to validate the existence of a "hard-wired" spiritual function in the human brain.

1) Take a group of ten highly spiritual or religious individuals from ten different religious orientations and submit them to SPECT (Single Positron Emission Computer Tomography) scans. See if the same part or parts of the brain show increased neural activity in every one of them.

1a) Conduct the same test on the same individuals, but instead of subjecting them to a SPECT scan, take blood from them to see if religious activity might prompt any noticeable difference in one's blood chemistry.

1b) Perform the same tests as above to a group of atheists all from different cultures and compare them to the results of the first group.

2) Take a group of one year olds. Perform SPECT scans on them. Have them undergo similar scans once every year, let's say, until they reach the age of twenty. Once a site has been identified that represents of the seat of spiritual cognition, look for changes to that site in each progressive scan taken on these test subjects. In this way, we might be able to chart the development of the spiritual function in the human brain.

2a) Pay special attention to those individuals who have undergone religious conversions. Compare the scan results of those who have undergone conversions, not only to their old scans (before they converted), but also to those who haven't converted at all.

3) Once a site has been identified in the brain through the use of SPECT scans as a possible seat of spiritual perceptions and/or experiences, study cases of individuals who have either had that part surgically removed or who have suffered some sort of damage to that area and see, to what degree, if any, this may have affected these individual's spiritual sensibilities. Such tests would confirm whether or not it is possible for humans to suffer from spiritual aphasias.

4) Compare and contrast the SPECT scans as well as the religious and spiritual propensities of identical twins separated at birth.

5) Compare and contrast the accounts of those who have had near-death experiences with those have been given the drug ketamine.

ENDNOTES

1 — William Keeton, <u>Biological Science</u> p.896

2 — Ibid A8

3 — Ibid p.65

4 — Ibid p.491

5 — Ibid p.492

6 — Ralph Linton, <u>The Science of Man in the World Crisis</u> p.123

7 — John Blacking, <u>How Musical is Man?</u> p.7

8 — Anthony Storr, <u>Music and the Mind</u> p.1

9 — Ibid p.29

10 — Ibid p.35

11 — Ivar Lissner, <u>Man, Magic and God</u> p.12

12 — Dr. Herbert Benson, <u>Timeless Healing</u> p.198

13 — E.O. Wilson, <u>On Human Nature</u> p.176

14 — Carl Jung, <u>Collected Works</u> (vol.9 Part 1) p.4-5

15 — Frieda Fordham, <u>An Introduction to Jung's Psychology</u> p.70

16 — Heobel. E and Frost. E, <u>Cultural and Social Anthropology</u> p.348

17 — Bronislaw Malinowski, "The Group and the Individual in Functional Analysis" (American Journal of Sociology 44 (May 1939):959)

18 — <u>Encyclopedia Britannica</u> (Fifteenth Edition, Macropaedia: Book 1) p.127

19 — Sigmund Freud, <u>Civilization and Its Discontents</u> p.25

20 — Ernest Becker, <u>Denial Of Death</u> p.17

21 — Ralph W. Hood Jr., Bernard Spilka, Bruce Hunsberger, & Richard Gorsuch, <u>The Psychology of Religion</u> p.153

22 — G. Zilboorg, "Fear of Death" Psychanalytic Quarterly 1943 12:465-467

23 — <u>Encyclopedia Britannica</u> (Macropaedia: Book 16) p.201

24 — Benson, <u>Timeless Healing</u> p.198

25 — S. Freud, <u>The Future of an Illusion</u> p.22

26 — Freud, <u>Civilization and its Discontents</u> p.20

27 — E.O. Wilson, <u>Sociobiology</u> p.2

28 — <u>Psychology of Religion</u>, p.161

29 — M. Ostow and B.A. Scharfstein, <u>The Need to Believe</u> p.23

30 — Karen Armstrong, <u>A History of God: The 4,000 year quest of Judaism, Christianity and Islam,</u> p.208

31 — Freud, <u>Civilization and Its Discontents</u> p.11

32 — Ibid p.12

33 — R.W. Hood, Jr., <u>Mysticism</u> p.285-297

34 — R.K. Forman, <u>The Problem of Pure Consciousness</u> p.8

35 — R.M. Bucke, Cosmic Consciousness: <u>A Study of the Evolution of the Human Mind</u> p.67

36 — M.M. Poloma and B.F. Pendleton, <u>Review of Religious Research</u> (1989) p.48

37 — Savage, Hoffman, Fadiman, Savage-1971

38 — J. Jaynes <u>The Origin of Consciousness in the Breakdown of the Bicameral Mind</u> p.360

39 — R.D. Laing: From Ralph Metner's <u>The Ecstatic Experience</u> p.15

40 — Wilson, Elgin, Vaughan, Wilber "Paradigms In Collision" from <u>Beyond Ego: Transpersonal Dimensions in Psychology</u> p.47

41 — <u>Beyond Ego</u> p.47

42 – Daniel Goleman "A Map for Inner Space" from Beyond Ego p.147
43 – C.D. Batson and W.L. Ventis, The Religious Experience p.98
44 – M. Pafford, Inglorious Wordsworths p.262
45 – R. Walsh, D. Elgin, F. Vaughan, K. Wilber, "Paradigms in Collision" from Beyond Ego p. 41
46 – R. Stark "A Taxonomy of Religious Experience" (Journal for the Scientific Study of Religion, 5, 1965) p.165-176
47 – W. James, Varieties of The Religious Experience p.315
48 – Woodruff (1993) Report: Electroencepholograph taken from Pastor Linton Pack, In T. Burton, "Serpent-Handling Believers" p.142-144
49 – Stanislav Grof, Realms of the Human Unconscious p.204
50 – J. Blofeld, The Tantric Mysticism of Tibet p.24
51 – Psychology of Religion p.229
52 – Dr. Raj Persaud, Financial Times (May 8/May 9, 1999) X Weekend FT
53 – Freud, Civilization and its Discontents p.13
54 – Soren Kierkegaard, Sickness Unto Death p.146
55 – V.S. Ramachandran, Phantoms in the Brain p.250
56 – Ibid p.252
57 – Ibid p.225
58 – Freud, Civilization and its Discontents p.14
59 – Norman O. Brown, Life Against Death p.159
60 – Freud, Civilization and its Discontents p.16
61 – Ibid p.21
62 – B.K. Anand, G.S. Chhina, B. Singh "Electroencephalography and Clinical Neurophysiology" 13, 1961 p.452-456
63 – Steven Rose, The Conscious Brain p.335
64 – Batson and Ventis, The Religious Experience p.98
65 – Robert Jesses "Entheogens: A brief history of their Spiritual Use" (Tricycle: Volume 6, Number One: Fall 1996) p.60
66 – Ibid p.62
67 – Benson, Timeless Healing p.157
68 – Ibid p.157
69 – W. James Varieties of the Religious Experience p.162
70 – Psychology of Religion p.279
71 – Ibid p.117
72 – S. De Sanctis, Religious Conversion: A Bio-Psychological Study p.67
73 – Psychology of Religion p.289
74 – Ibid p.280
75 – Pratt, J.B., The Religious Consciousness: A Psychological Study p.113
76 – Ostow and Scharfstein, The Need to Believe p.102
77 – Psychology of Religion p.279
78 – Ibid p.399
79 – Fenwick, Peter, M.D. The Truth of the Light p.13
80 – K.L.R. Jansen, M.D. Using Ketamine to Induce the Near-Death Experience p.64
81 – Ibid, p.73
82 – E.O. Wilson, Sociobiology p.287
83 – Psychology of Religion, p.19
84 – S. Aurobindo, The Future Evolution of Man p.27
85 – A.N. Whitehead, Religion in the Making p.37

BIBLIOGRAPHY

American Psychiatric Association (1994) Diagnostic and Statistical Manual of mental disorders (4th ed.) Washington, D.C.: Author

Armstrong, Karen (1993). *A History of God: The 4,000 year quest of Judaism, Christianity and Islam.* Knopfs.

Aurobindo, S. (1963). *The Future Evolution of Man.* All India Press.

Bateson, G. (1979). *Mind and Nature.* E.P. Dutton.

Batson, C.D. and Ventis, W.L. (1982). *The Religious Experience.* Oxford University Press.

Becker, E. (1973). *Denial Of Death.* The Free Press.

Benedict, R. (1989). *Patterns Of Culture.* Houston Mifflin.

Benson, H. (1996). *Timeless Healing.* Scribner.

Blacking, J. (1976). *How Musical is Man?.* Faber & Faber.

Bootzin, R., Acocella, J.R., Alloy, L.B. (1993). *Abnormal Psychology.* McGraw-Hill Inc.

Brown, N.O. (1959). *Life Against Death.* Vintage Books.

Bucke, R.M. (1961). *Cosmic Consciousness: A Study of the Evolution of the Human Mind.* University Books.

Campbell, J. (1990). *Transformations of Myth Through Time.* Harper and Row.

Cavendish, R. (1980). *Mythology.* Rizzoli International Publications Inc.

Chance, M.R.A. (1962). *Social Behavior and Primate Evolution.* In M.F. Ashley Montagu, ed. Culture and the Evolution of Man. Oxford University.

Choron, J. (1963). *Death and Western Thought.* Collier Books.

Clark, W.H. (1969). *Chemical Ecstasy: Psychadelic Drugs and Religion.* Sheed and Ward.

Coren, S., Porac, C., and Ward. L,. (1978). *Sensation and Perception.* New York; Academic Press

Cohen, D. (1991). *The Circle of Life: Rituals From the Human Family Album.* Harper San Francisco

Davidson, J. And Davidson, R. (1979). *The Psychology of Consciousness.* Plenum.

Dobzhansky, T. (1963). *Anthropology and the Natural Sciences – the Problem of Human Evolution.* Current Anthropology, 4:138, 146-148.

Durant, W. (1961). *The Story of Philosophy.* Washington Square Press

Eccles, J.C. (1953) *The Neurophysiological Basis of Mind.* Clarendon Press.

Eliade, M. (1987). *The Encyclopedia of Religion.* MacMillan.

Eliade, M. (1959). *The Sacred and The Profane: The Nature of Religion.* Harcourt Brace Jovanovich

Fordham, F. (1953). *An Introduction to Jung's Psychology.* Penguin Books.

Forman, R.K. (1980). *The Problem of Pure Consciousness.* Oxford Univ. Press.

Freud, S. (1962). *Civilization and Its Discontents.* W.W. Norton and Co. Inc.

Freud, S. (1926). *Inhibitions, Symptoms and Anxiety.* Translated by Strachey, A. New York: Norton

Freud, S. (1927) *The Future of an Illusion.* J. Strachey, Trans. Norton.

Fromm, E. (1955). *The Sane Society.* Fawcett Publications.

179

Foucalt, M (1965). *Madness and Civilization*. New York: Random House

Furst, P. (1972). *Flesh of the Gods:The Ritual Use of Hallucinogens*.Praeger Publishers

Gallup, G.G., Jr. (1968). Mirror-image stimulation. *Psychological Bulletin*. 70: 782-793

Gallup, G.G.,Jr. (1970). Chimpanzees: Self-recognition. Science 167: 86-87

Geschwind, N., (1972). Language and the brain. *Scientific American* 226: 76-83

Goleman, D. And Davison, R. (1979). *Consciousness: Brain, States of Awareness and Mysticism*. Harper and Row.

Greyson, B. (1983) The Psychodynamics of a Near-Death Experience. Journal of Nervous and Mental Disease. 376-380

Grof, S. (1975). *Realms of the Human Unconscious*. The Viking Press.

Hamilton, W.D. (1964). *The Genetical Theory of Social Behavior*. I,II. Journal of Theroetical Biology, 7:1-52.

Hamilton, W.D. (1970). *Selfish and Spiteful Behavior in an Evolutionary Model*. 228:1218-1220.

Heelas, P. (1985). *Social Anthropology and the Psychology of Religion*. Elsmsford, NY. Pergamon Press

Heobel. E and Frost. E. (1976). *Cultural and Social Anthropology*. McGraw-Hill Book Company.

Hill, P.C. (1995) *Affective Theory and Religious Experience*. Birmingham AL; Religious Education Press

Hood, R.W. Jr., Spilka, B., Hunsberger, B. & Gorsuch, R. (1996) *The Psychology of Religion*. The Guilford Press

Jansen, K.L.R. (1991) Transcendental Explanations and the Near-Death Experience. Lancet. 337, 207-243

Jansen, K.L.R. (1993) Non-Medical use of Ketamine. British Medical Journal. 298, 4708-4709

Jansen, K.L.R. (1995) Neuroscience, Ketamine and the Near-Death Experience; The Role of HGlutamate and the NMDA Receptor in the Near-Death Experience. Routledge, N.Y.

Jansen, K.L.R. (1996) The Ketamine Model of the Near-Death Experience: A Central Role for the NMDA Receptor. In Journal of Near_Death Studies (Ed. B.Greyson)

Janssen, J., de Hart, J., & Gerhadts, M. (1994). Images of God in Adolescence. *International Journal fthe the Psychology of Religion*. 4, 105-121

Jung, C.G. (1958) *The Undiscovered Self*. Mentor.

Jung, C.G. (1967-1978). *Man and His Symbols, Psychology of Religion, and Symbolic Life*. Translated by R.F.C. Hull. Bollingen Series XX. Princeton Univ. Press.

Jung, C.G. (1976). *The Portable Jung*. Edited by Joseph Campbell. Translated by R.F.C. Hull. Penguin Books.

James, W. (1902). *Varieties of The Religious Experience*. Collier Books.

Jesses, R. (1996). *Entheogens: A brief history of their Spiritual Use*. Tricycle: Volume 6, Number One.

Jones, W.L. (1937). *A Psychological Study of Conversion*. London; Epworth

Katz, S.T. (1992). *Mysticism and Language*. New York; Oxford University Press

Kierkegaard, S. (1974). *Sickness Unto Death*. Princeton Univ. Press.

Keeton, William (1980). *Biological Science*. W.W. Norton and Company Inc.

Kohlberg, L. (1984) *Essays on Moral Development: Vol. 2. The Psychology of Moral Development: The Nature and Validity of Moral Stages*. San Francisco; Harper and Row

Kolb, B., & Whishaw, I.Q. (1990). *Fundamentals of Human Neuropsychology*. (3rd ed.). New York; W.H. Freeman

Leary, T.F. (1983) Flashbacks, An Autobiography. J.P. Tarcher. L.A. p.375

Lee Hotz, Robert, *Seeking the Biological Origins of Spirituality;* Los Angeles Times; April, 26 1998

Le Doux, Joseph. (1994). Emotion, Memory and the Brain. *Scientific American* 270: 32-39

Le Doux, Joseph (1996). *The Emotional Brain*. Simon & Schuster

Leiman, A. and Rosenzweig, M. (1982) *Physiological Psychology*. D.C. Heath and Company.

MacKenzie, N. (1965) *Dreams and Dreaming*. London: Aldus Books

Malinowski, B. (1939) *The Group and the Individual in Functional Analysis,* American Journal of Sociology 959,44.

May, R. (1950). *The Meaning of Anxiety*. Ronald Press.

Morris, D. (1969). *The Naked Ape*. Jonathan Cape

Morris, D. (1970). *The Human Zoo*. Jonathan Cape

Noelle C. David (1998). Searching for God in the Machine; *Free Inquiry*; Summer

Ostow, M. and Scharfstein, B.A. (1953). T*he Need to Believe*. International Univ. Press.

Parrinder, G. (1971). *World Religions: From Ancient History to the Present*. Facts on File Publications.

Persaud, Raj, *God's in your Cranial Lobes,* Financial Times, May 8/May 9 1999

Persinger, Michael (1987). *Neuropsychological Bases of Belief.* New York; Praeger

Pfeiffer, F. (1977). *The Emergence of Man*. McGraw-Hill.

Pratt, J.B. (1920). *The Religious Consciousness: A Psychological Study*. New York: MacMillan

Prince, R.H., (1992). *Religious Experience and Psychopathology*. New York: Oxford University Press

Pruyser, P.W. (1968). *A Dynamic Psychology of Religion*. New York: Harper and Row

Pugh, G. (1977). *The Biological Origin of Human Values*. Basic Books.

Ramachandran, V.S. (1998). *Phantoms in the Brain: Probing the Mysteries of the Human Mind*. New York: William Morrow & Co., Inc

Rank, O. (1961). *Psychology and the Soul*. Perpetua Books.

Reynolds, F.E. and Waugh, E.H. (1977). *Religious Encounters with Death: Insights from the History and Anthropology of Religion*. Pennsylvania State Univ. Press.

Richmond, P.G. (1970). *An Introduction to Piaget*. Routledge and Kegan Paul

Rose, Steven (1976). *The Conscious Brain*. Vintage Books.

Rosenbleuth, A. (1970). *Mind and Brain*. The M.I.T. Press.

Sacks, O. (1973). *Awakenings*. Duckworth.

Sputz, R. (1989) I Never Met a Reality I didn't Like: A Report on 'Vitamin K'. High Times. October 1989, 64-82

Stark, R. (1965). *A Taxonomy of Religious Experience*. Journal for the Scientific Study of Religion, 5.

Storr, A. (1992). *Music and the Mind*. Ballantin.

Tillich, Paul (1952). *The Courage to Be*. Yale University Press

Turnbull, C. (1983). *The Human Cycle*. Simon and Schuster.

Turner, V. (1969). *The Ritual Process*. Aldine Publishing Company.

Van Gennep, A. (1960). *The Rites of Passage*. The Univ. Of Chicago Press.

Walsh, R and Vaughan, F. (1980). *Beyond Ego: Transpersonal Dimensions in Psychology*. J.P. Tarcher Inc.

Whitehead, A.N. (1926). *Religion in the Making*. Macmillan.

Wilson, E.O. (1976). *On Human Nature*. Bantam Books.

Wilson, E.O. (1980). *Sociobiology*. London: The Belknap Press of Harvard University Press.

Young, J.Z. (1964). *A Model of the Brain*. Clarendon Press.

Zilboorg, G. (1943). *Fear of Death*. Psychanalytic Quarterly, 12:465- 467.